Labor Standards and Metal Mining

A Study of the Effects of
The Fair Labor Standards Act of 1938
on the Non-ferrous Metal Mining Industry

by TELL ERTL and THOMAS T. READ

Submitted in partial fulfillment of the requirements
for the degree of Doctor of Philosophy, in
the Faculty of Pure Science,
Columbia University

KING'S CROWN PRESS
at Columbia University, New York
1947

Copyright 1947 by

TELL ERTL

Printed in the United States of America

KING'S CROWN PRESS
is a division of Columbia University Press organized for the purpose of making certain scholarly material available at minimum cost. Toward that end, the publishers have adopted every reasonable economy except such as would interfere with a legible format. The work is presented substantially as submitted by the author, without the usual editorial attention of Columbia University Press.

Typed by Eloise Beckwith

Printed by Edwards Brothers
Ann Arbor, Mich.

CONTENTS

INTRODUCTION	1
I. THE NON-FERROUS METAL MINING INDUSTRY	3
A. Location of the Industry	3
B. Operation of the Industry	7
C. Size of the Industry	13
D. Conclusion	14
II. HISTORICAL BACKGROUND OF THE FAIR LABOR STANDARDS ACT	15
A. The Viewpoint of Congress	15
B. The Viewpoint of the Mining Industry	18
C. Conclusions	22
III. EFFECTS OF THE ACT	23
A. The First Period	23
1. Working Schedule Adjustments	24
2. Cost Changes	25
3. The Administration of the Act	28
B. The Second Period	28
1. Work Defined	31
2. The Effect of the Second Period	40
C. The Third Period	40
D. Summary	42
IV. INVESTIGATION OF METHODS FOR SURMOUNTING INCREASED COSTS	45
A. Improvement of Labor Productivity	45
B. The Administration of Working Time	47
1. Reduction in the Proportion of Non-Productive Time	49
2. The Use of Swing Shifts and Staggered Schedules	52
3. The Operation of Multiple Shifts	73
C. Summary	76
CONCLUSION	77
NOTES	79

INTRODUCTION

"I am under the impression that no other industry has presented more complicated problems of compliance to us than metal mining. I think the real explanation [for these compliance problems] is that the industry really is different. In all the country there is no other quite like metal mining."

This statement was made by General Philip C. Fleming, Administrator, Wage and Hour Division, United States Department of Labor, in a paper presented before the Annual Metal Mining Convention, Western Division, American Mining Congress, Colorado Springs, Colorado, September 16 to 19, 1940.[1] The industry addressed was the non-ferrous metal mining industry, consisting primarily of producers of copper, zinc, lead, silver, and gold. What follows here is a study of the non-ferrous metal mining industry and the effects of the Fair Labor Standards Act on it.

Why compliance with the Act was such a problem to the non-ferrous metal mining industry is not immediately evident. The Act requires every industry which operates in interstate commerce (1) to pay minimum wages of at least forty cents an hour, (2) to employ no child labor, and (3) to pay labor one and one-half times the normal rate for all hours that are worked above forty in any week.[2] Since the entire non-ferrous metal mining industry has been adjudged to operate in interstate commerce, it therefore must abide by the Act.

Requirement (1) fixes a minimum pay rate of forty cents an hour. The non-ferrous metal mining industry always has had a high wage scale. In 1939 the copper mining industry paid an average of sixty-seven cents an hour, the lead and zinc mining industry paid sixty-two cents an hour, gold sixty-seven, and silver sixty-six cents an hour.[3] Only unskilled work in such low-wage or depressed areas as Northern Michigan, parts of New Mexico, and in some portions of the Southeastern states was paid at rates less than forty cents an hour during 1938 and 1939.[4] Julian D. Conover, Secretary, American Mining Congress, stated to the joint labor committees of the Senate and the House on June 15, 1937, that "wages of mine employees are substantially in excess of the statutory minimum contemplated,"[5] and Edward H. Snyder, General Manager of Combined Metals Reduction Company, corroborated Mr. Conover when he said before the Annual Metal Mining Convention, on September 10, 1937, "The minimum wages now being paid in the mining industry range from 25 to 65 percent higher than the highest minimum wage that could be established by the board."[6] It is obvious, then, that the industry has not had any difficulty complying with the minimum wage provisions of the Act.

Mr. Conover further stated to the committee that child labor is not a problem in the mining industry.[5] More than a century ago, in 1842, the English Parliament passed an act prohibiting the employment of boys under 10, and of all girls and women, in mine pits. By October, 1929, thirty states, including most of those in which mining is an important industry, had prohibited the employment of boys in mines before they had reached the age of 16, and five states, among them Arizona and New Mexico, had a still higher minimum age.

The total number of children below the age of sixteen who were reported at work in mines, oil wells, and quarries in 1920 was 7191; in 1930, 1184; and in 1940, 219.[7,8,9] Most of the children were employed in coal mines and only a few in non-ferrous metal mines.[7] Such very important mining states as Arizona, Montana, Nevada, and Utah had respectively only 13, 15, 5, and 7 children employed in the extraction of minerals in 1920. Certainly, the non-ferrous metal mining industry has not had any difficulty complying with the minimum age provisions of the Act.

The non-ferrous metal mining industry, therefore, found compliance with the minimum wage and the minimum age provisions of the Fair Labor Standards Act no hardship. Consequently, the "complicated problems of compliance" to which General Fleming refers must arise from the provision of the Act that requires the payment of one and one-half times the regular hourly rate for all hours that are worked above forty in any week. In the following pages the objective is to show how the hours provision affects the non-ferrous metal mining industry and what measures the industry has taken in the past, and may take in the future, to overcome the disturbing effects of the Act.

Chapter I

THE NON-FERROUS METAL MINING INDUSTRY

The non-ferrous metal mining industry includes the mining of copper, zinc, lead, gold, and silver. The importance of the industry and the value of the non-ferrous metals to man can be demonstrated by the uses to which they are put.

Copper,[1] the chief metal of the group, conducts most of the electricity used in the world. Other important uses of copper are for ammunition and in the manufacture of brass and bronze. Zinc and lead rank next to copper and are of equal importance to man. Zinc is used as a metal mainly for galvanizing steel sheets, in the manufacture of brass, and for die-casting intricate shapes, such as automobile door handles. Lead is used to a large extent for storage batteries and cable coverings. Large quantities of both zinc and lead compounds are used to make up the pigments, white lead, red lead, litharge, zinc oxide, and lithopone, of the paint industry. Gold and silver are primarily bases for the world's monetary systems, though considerable amounts are used for jewelry and in industry.

A. SITUATION OF THE INDUSTRY

All non-ferrous metals are produced from ores. An ore is a mineral or an aggregate of minerals from which one or more metals can be extracted at a profit on a commercial scale. Ore deposits vary in size from "pockets" that contain a few tons of mineral to bodies from which over a billion tons of metal-bearing rock will be extracted before ultimate exhaustion. The orebody of the Utah Copper Company in the Oquirrh Range, southwest of Salt Lake City, has been estimated to contain one billion tons of copper ore, and the Chuquicamata orebody in the Andes of Northern Chile is known to be even larger.[2]

At the present stage of technological advancement, the mining of metals occurs in the upper ten thousand feet of the earth's crust. Clarke and Washington have estimated that the upper ten miles of strata contain an average of 0.01% copper;

0.002% lead; 0.0004% zinc; 0.000,004% silver; and 0.000,000,1% gold.³ Before these metals can be mined at a profit, their minimum content in the mined rock must be in the order of 1% for copper, 4% for lead, 3% for zinc, 0.0001% for silver, and 0.000,001% in the case of gold.⁴

If these metals were uniformly distributed throughout the earth, they probably could not be exploited. Since they actually occur in workable concentrations as mineral deposits, they have been mined since man first used weapons, tools, or ornaments.

Metallic mineral deposits were formed during all of geological time, generally during periods of, and in areas of, igneous activity.⁵ Igneous activity is associated with tectonic movement or mountain building. Consequently, the principal metallic mineral deposits of the world are typically found in mountainous regions, or in the eroded remnants of former mountains.

A saying among mining engineers is that no two mineral deposits are exactly alike. Some are regular in width and value, most are not. Some are thin and tabular, others thick. Some lie flat, some stand vertically; the majority are anywhere between these extremes. Some contain predominantly one metal, many contain two to five or even more metals. Nature has produced an endless variety of conditions so that each orebody resembles others in some aspects, but always differs in details that must be understood and coped with in order to mine it successfully.

A mineral deposit, when discovered, is as big as it will ever be; it contains as much metal as it will ever contain, concentrated to its ultimate degree of purity, limited by its present boundaries. The miner cannot leave his orebody and expect it to grow larger as can the lumberman his tree.⁶ T. A. Rickard states:

> The suggestion has been made that perhaps the idea behind the interdiction of mining by the [Roman] Senate was to give the mineral deposit a chance to grow, and thus become more productive in the future, as land is allowed to lay fallow for the sake of a better crop the next year. It was common belief among the ancients that metal grew in mines after they were abandoned. A pupil of Aristotle, Theophrastus, the Roman, and Pliny, believed this myth. Livingstone records that Negroid natives of Manicaland bury a particle of gold in the belief that it is the seed from which a crop will be obtained in due season. We know now

that mineral deposits increase by deposition from the chemical solutions that circulate underground, but we realize that the concentration of ore at any given place is due to a process so slow as rarely to be effective within the lifetime of a man.

The miner cannot enlarge his orebody to meet increased demand as the manufacturer can enlarge his plant. The miner cannot extract his ores, replant part, and return at a later date to reap again as he has sown. A mineral resource is an exhaustible resource, and once exhausted is no more.

Since orebodies are fixed and cannot be moved except as they are mined, the miner must operate his property with all the inherent disadvantages of situation which the mineral deposit originally possesses. Most of the non-ferrous metal production of the United States is from the states west of the Mississippi River and far from consumers in the industrial East. Only a minor portion is produced east of the Mississippi. The mining districts west of the Rockies are often in arid, mountainous regions, and far from abundant sources of food, water, timber, and power. Some are at high altitudes, or in areas of excessive snowfall. Mines, at the time of discovery, are usually long distances from roads, railroads, and other means of travel and communication. Many mines are far from smelters and refineries; most are distant from metal markets.

An index of this relation can be found by referring to Table 1-1, which shows the population per square mile of the states that produced most of the copper, zinc, lead, gold, and silver in 1939. The chief non-ferrous metal mining states are areas with a low density of population. The mining districts are often in the more isolated portions of these relatively unpopulated areas. A questionnaire circulated by the Wage and Hour Division produced the unpublished results shown in Table 1-2. This table illustrates how far much of the mining industry operates from the normal stream of American life. Many medium-sized mines employing one hundred men or more are scores of miles from towns with populations of a few hundred or a few thousand. Such small towns can offer little more than a saloon or a movie for the social life of the miners. Even the small towns are sometimes inaccessible to the mine employees in the winter months (see plate 3-1). Since the mining "camp" is usually built and owned by the mining company, and since it often has no other reason for existence than to house the employees of

TABLE 1-1

State	Value of Cu, Zn, Pb, Au, Ag, Production in 1939 in Millions of Dollars*	Percentage of Value by Metals*					Population per square Mile†
		Cu	Au	Pb	Ag	Zn	
Arizona	$72.6	75	15	1	8	1	4.4
Utah	62.7	57	15	10	12	6	6.7
California	52.9	2	95	t	3	t	44.1
Montana	40.9	44	23	4	15	9	3.8
Nevada	30.5	46	41	1	10	2	1.0
Idaho	29.8	2	14	29	39	16	6.3
Alaska	23.9	t	99	t	1	t	0.1
Colorado	22.3	12	58	3	26	1	10.8
South Dakota	21.8	0	100	0	t	0	8.4
Oklahoma	17.2	0	0	15	0	85	33.7
Missouri	16.4	0	0	89	1	10	54.6
New Mexico	15.4	62	9	3	6	20	4.4
New Jersey	11.5	0	0	0	0	100	555.1
Michigan	9.2	99	0	0	1	0	92.2
Kansas	8.5	0	0	15	0	85	21.9
Washington	6.9	28	46	7	4	15	23.3
New York	3.8	0	0	?	1	99	281.2
Oregon	3.4	t	98	t	2	0	11.3
Tennessee	?	?	?	?	?	?	69.5
Pennsylvania	?	?	?	?	?	?	219.8

t Represents production less than 0.5 percent.
* Minerals Yearbook, Review of 1940.
† World Almanac 1942.

the mining company, no more houses than are necessary to accommodate the working force are built. Table 1-2 shows that only fifteen of the thirty-eight mines which reported on the question had surplus housing facilities. Plates 1-1, 1-2, and 1-3 show typical mining areas in the states of Washington, Nevada, and South Dakota. The photographs show under what physical handicaps of situation the non-ferrous metal mining industry often labors.

B. OPERATION OF THE INDUSTRY

A mining enterprise performs three operations on an orebody: exploration, development, and extraction or mining. Exploration includes the procedures that lead to the discovery of a new orebody or the extension of a known one. By the use of the methods and equipment of modern prospecting the orebody is explored to determine its extent and richness. If it appears that enough mineral-bearing rock of a grade that will permit profitable mining exists, the deposit will next undergo development.

The preparation of a mineral deposit for extraction of the contained ore is called development. Development work includes the excavation of all openings required for transportation of men, materials, and rock, and those required for ventilation and drainage. As development reveals more information concerning the size, shape, and grade of the orebody, it may also be considered continued exploration; and as some valuable minerals may be extracted in the excavation of development openings, it may also be considered, in part, mining.

The economic purpose of exploration and development is to make its later extraction possible. If no ore is discovered during exploration, or if development reveals conditions that will make extraction unprofitable, the money spent can never be recovered. Mine openings cannot be utilized for any other purpose. Mining, therefore, is the removal of the valuable metalliferous rock, the ore, which was discovered by exploration along openings made during development.

When the mining operations extract the ore, it is generally put through an additional step before it is delivered to another industry. The ore is beneficiated or milled. Beneficiation is the process by which most of the desirable metallic minerals are separated from the valueless remainder of the ore. The economic justification of beneficiating ore prior to smelting is to reduce losses of metal at the smelter and to reduce the smelting cost by sending a smaller quantity of

TABLE 1-2*: Metalliferous Mining:
Illustrations of data which bear on the problem of isolation

Type of Mine	Principal Metal Mined	Number of Employees Mine and Mill 1939	Distance from Nearest Town (Miles)
Underground	Gold	65	11
	Gold	85	25
	Gold	272	4
	Gold	7	9
	Silver	30	3
	Gold	40	7
	Gold	7	11
	Lead-Silver	34	1
	Gold	66	48
	Gold	117	6
	Gold	93	2
	Gold	71	40
	Gold	143	2
	Gold	47	9
	Silver	58	30
	Gold	5	12
	Gold	76	12
	Silver	137	9
	Gold	37	60
	Silver	40	21
	Gold	220	1
	Gold	190	1
	Gold	11	8
	Gold	5	85
	Copper	36	32
	Copper	15	18
	Copper	318	5
	Copper	459	57
	Copper	550	26
	Miscellaneous	81	6
	Miscellaneous	8	6
Open Pit and Underground	Miscellaneous	208	68
Open Pit	Gold	34	14
Placer	Gold	21	2
	Gold	2	3
	Gold	20	1
	Gold	11	2
	Gold	14	14
	Gold	13	7
	Gold	10	7
	Gold	31	6

a. Answers to question 7D: "For approximately how many days during the win of 1938-1939 was the mine not accessible by automobile?"

b. Answers to question 7A: "Please estimate the number of additional employe who could be housed in living quarters now available reasonably near the mine."

Population of Nearest Town	Days Inaccessible by Auto in Winter of 1938-39[a]	Number of Additional Employees Who Could be Housed[b]
100	None	8
5,000	None	None
1,500	All	None
1,500	120[c]	None
400	5	20
110	None	12
642	None	----
125	None	20
651	210	None
400	None	20
600	6	20
12,000	25	None
250	None	10
800	None	12
9,000	None	None
350	5	4
350	None	12
1,200	None	None
7,000	25	75
1,040	10	None
300	1	None
164	2	None
175	None	None
442	All	----
2,000	10	None
1,100	None	None
300	10	None
1,400	All	None
1,400	80	80[d]
50	None	None
50	90	10
3,700	10	None
30	196	None
50	None	----
100	None	None
75	None	None
50	60	4
300	90	6
100	None	None
300	None	None
600	None	None

c. Accessible by mule pack only on 120 days.
d. Accommodations for single men only.
* From Wage and Hour Division Report.

Plate 1-1.

A large mine in the Cascade Mountains, Washington, which can only be reached from the nearest town after a 45-mile trip by boat and eleven miles by automobile.

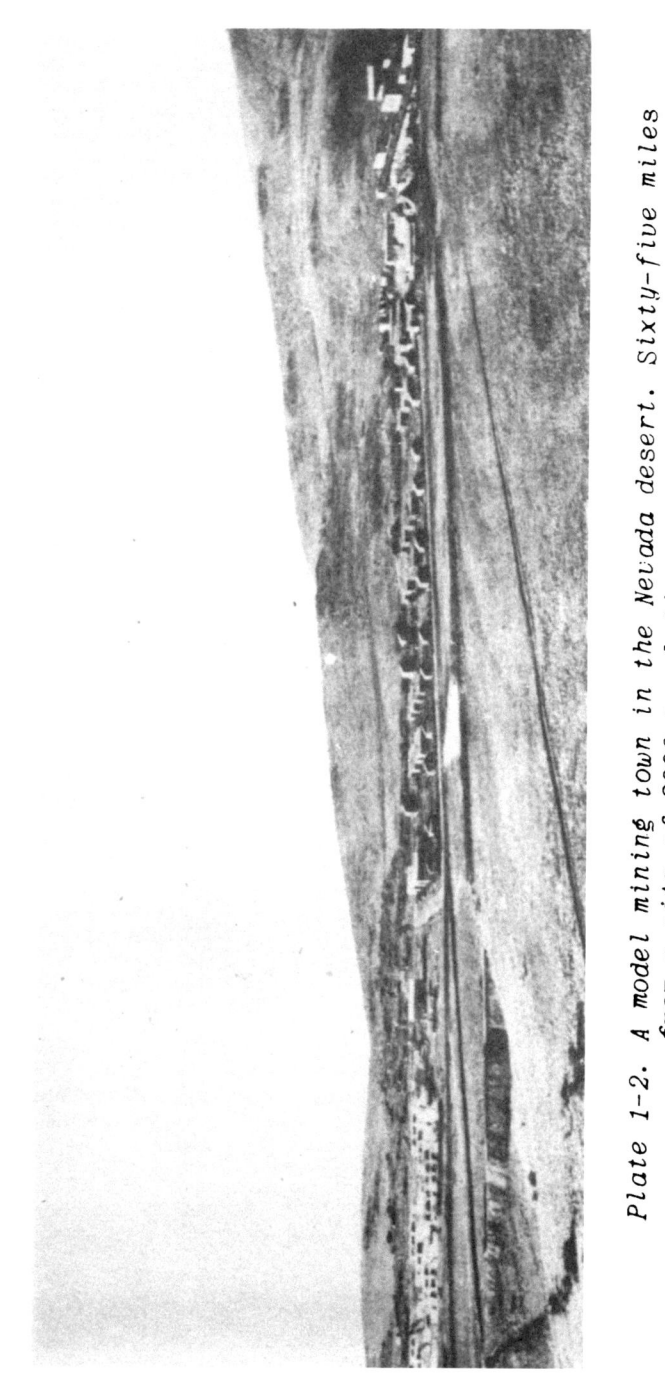

Plate 1-2. A model mining town in the Nevada desert. Sixty-five miles from a city of 3000 population.

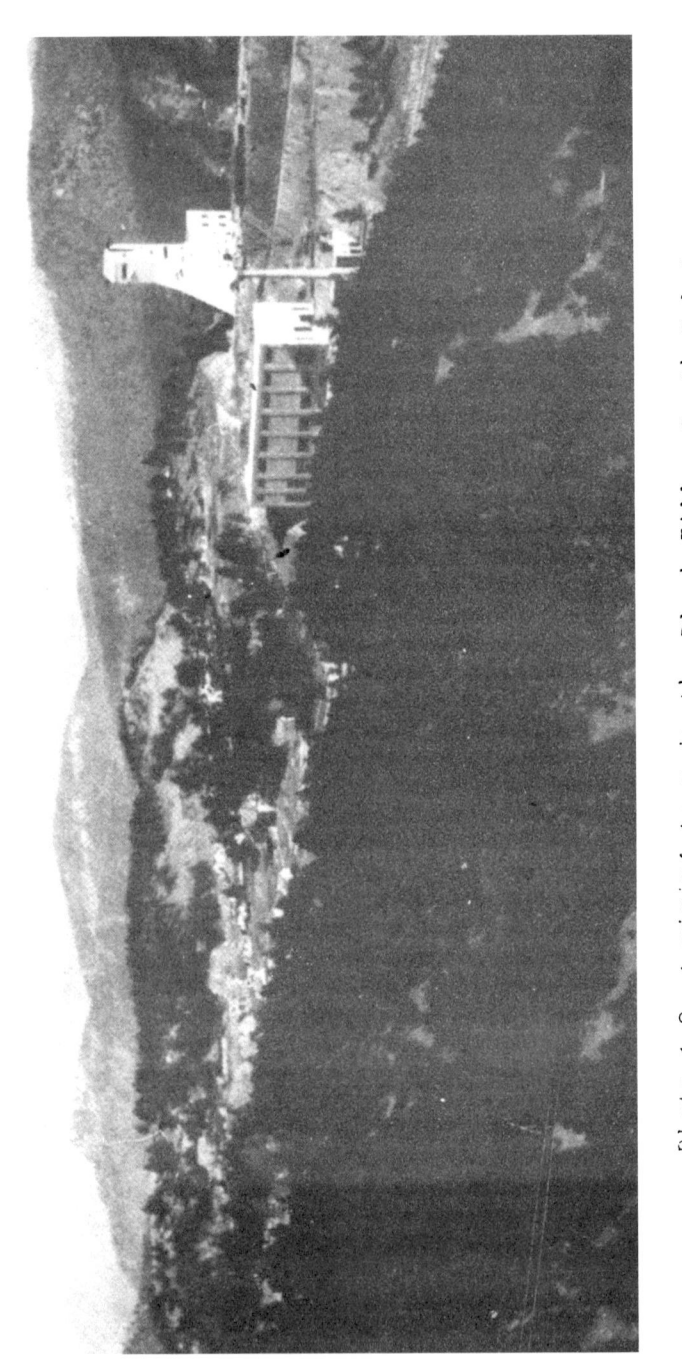

Plate 1-3. A mining town in the Black Hills, South Dakota.

material to the smelter.[7] If the mineral is milled at the mine, much less material need be transported at expensive haulage rates; therefore, most mills are built near the mine.[8] For this reason the effect of the Fair Labor Standards Act on the milling phase of the copper, zinc, lead, gold, and silver mining industry also will be considered.

C. SIZE OF THE INDUSTRY

The non-ferrous metal mining industry of the United States is the largest in the world. During 1940 the United States was the leading producer of copper, zinc, and lead; second in the production of silver; and fourth in the production of gold.[9] Over one-third of the world's copper and zinc, one-fourth of the lead and silver, and one-ninth of the gold mined in 1940 was produced in the United States.

The production came from a large number of orebodies of various sizes, shapes, positions, and grade. Relatively few orebodies have definite workable boundaries. In most cases the orebody is bounded by marginal mineralized material that is not rich enough to be mined at the then prevailing metal price. For example, a copper mine operator may determine that he can extract all material of which the copper content is above one percent, at a copper price of ten cents per pound. His deposit may contain as much of nine-tenths percent copper-bearing material as the tonnage of ore with a grade greater than one percent. If the price of copper should rise above eleven cents per pound, with no rise in mining cost, the amount of minable ore in the mine is doubled by the addition of the tonnage of nine-tenths percent copper-bearing material to the ore reserve.

It is not unusual in the mining industry for a small rise in price of the produced metal to cause a large increase in the ore reserve. Conversely, it is not unusual for a drop in price to reduce the ore reserve so much that continued profitable operation becomes impossible. The same effects result from increases or decreases in the cost of production although the metal price remains stationary. The mining industry, then, is affected more by small increases or decreases in the price of the product, or by small changes in the cost of operating, than are most industries.

The industry operates on an exhaustible reserve that is limited in the quantity of metal contained. Sensible economics dictates that extraction should continue only when the metal price is high and the cost of production not excessive, so that a high profit can be earned. If production is continued without profit, the mine is exhausted without return. The size of the non-ferrous metal mining industry, then, depends in a larger measure than most industries on the price of the product and its costs of production.

D. CONCLUSION

The non-ferrous metal mining industry "really is different" from other industries, to quote again from the statement made by General Fleming (p. 1). The situation of the industry is often remote, in mountainous or desert regions, possibly inaccessible some parts of the year. The towns adjacent to the mines are mining "camps" and usually exist only so long as the mine operates. Each mine is an impermanent enterprise, of relatively short life, since it operates on an exhaustible resource. The mine operator has no control over, and in most cases only partial knowledge of, the extent of his reserves. The extent of each reserve is changeable and varies greatly with cost of production and metal price. Many mines are relatively small operations and, except for a few open-pit mines, the work cannot be systematized as in factories. The extent to which, and the manner in which, the mining industry differs from other industries is the basic reason for the compliance difficulties with which the industry has been faced.

Chapter II

HISTORICAL BACKGROUND OF THE FAIR LABOR STANDARDS ACT

The historical background of this Act needs to be considered from two viewpoints; that of the Congress, which presumably represents the viewpoint of the general public, and that of the mining industry.

A. THE VIEWPOINT OF CONGRESS

The roots of the Fair Labor Standards Act are deep in a movement that extends throughout the period of industrial history. Hour legislation has been the generally accepted state practice for many years.[1] As early as 1898 the U. S. Supreme Court approved a Utah statute regulating the hours of labor for men working in mines. That decision laid the foundation for future hour laws when Mr. Justice Brown, speaking for the Court, said in part (Holder v. Hardy, 169 U. S. 366, 397): "But the fact that both parties are of full age and competent to contract does not necessarily deprive the state of the power to interfere where the parties do not stand upon an equality or where the public health demands that one party to a contract shall be protected against himself."

By January 1, 1939, forty-four states had limited to some degree the hours of labor for women, and five states likewise included men in their general laws; about a dozen states limited hours for men in specific industries. All of the western states, except New Mexico and Washington, and the central mining states (Kansas, Missouri, and Oklahoma) had passed laws forbidding a miner to work underground more than eight hours a day. Michigan limited the hours to ten hours a day. None of these states attempted to limit the days per week, or the hours per week below fifty-six (seventy in Michigan) that could legally be worked by men underground.

Since 1840, when President Van Buren by executive order stipulated the 10-hour day in government navy yards,[2] the Federal Government as employer has led industry in hour reduction. As early as 1868, when the 8-hour day in industry was little more than a dream, Congress decreed those hours for any workers employed on government contracts.[3] That law, though not entirely effective, was continued and by 1936 further congressional action had provided regulations to cover most possible cases.

Because excessive hours were recognized as a contributing factor in railroad accidents, both state and federal legislation was passed to cover the situation. By a law applicable to workers on interstate railways, Congress in 1907 set sixteen as the maximum number of hours to be worked in one day, and provided for adequate rest periods.[4] In 1916, at the insistence of President Wilson, Congress provided for a basic 8-hour day for railroad trainmen.[5] However, the Child Labor Act, passed by Congress in 1916, was declared unconstitutional by the U.S. Supreme Court in 1918, Justice Holmes dissenting.

By the time of the "New Deal," regulation of hours by government was a principle accepted in general by legislatures and courts alike. Most states had 10-hour laws for women and were considering regulations providing for shorter hours for both men and women. The Federal Government had likewise regulated in those fields open to it.

The direct history of the Act probably begins with the insistence of President F. D. Roosevelt on the labor standards in the National Industrial Recovery Act.[6] After this Act was declared unconstitutional by the Supreme Court in the Schecter case, President Roosevelt often deplored the abandonment of the labor provisions.[7] Nevertheless, in 1936, in the case of Morehead and People ex rel. Tipaldo (298 U.S. 587), the Supreme Court again declared that wage and hour legislation was unconstitutional.

The Democratic Party continued to urge some type of national action to eliminate substandard working conditions, and talk was heard of a possible constitutional amendment. In his message to Congress on January 6, 1937, however, President Roosevelt stated his general opposition to immediate amendment to the Constitution, and asked instead for an "enlightened view" on social legislation from the judiciary. Reports of President Roosevelt's first press conference for the year 1937 indicate that plans were being formulated to "do something" about minimum wages as well as judicial opposition to his program.[8] The coincidence of his statements on these two matters serves to illustrate the close tie between federal labor standards legislation and the President's plan for reorganization of the Supreme Court. On February 5,

1937, President Roosevelt announced his plan for reorganizing the federal judiciary, generally known as the "Court packing" plan.

Soon afterwards, the Supreme Court reversed its previous stand on minimum wage legislation (West Coast Hotel Co. v. Parrish, 300 U. S. 379), upheld the Railway Labor Act (Virginian Ry. v. System Federation No. 40, 300 U. S. 515), and the National Labor Relations Act (National Labor Relations Board v. Jones and Laughlin Steel Corporation, 310 U. S. 1). These decisions served to take considerable pressure off the drive to enact the Court-packing plan, and spurred the administration supporters to a decision to introduce a wage and hour bill.

On May 24, 1937, nearly identical bills, S. 2475, and H. R. 2700, were introduced by Senator Black and Representative Connery. Not until one year, one month, and one day had elapsed did the measures finally reach the President's desk for signature, and then only after having undergone amendment after amendment until practically the only point in common with the original bill was the legislative number. Few legislative enactments in American history have had such a stormy career and assumed so many different aspects within a comparatively short period of time as has this Act.

The passage of the Fair Labor Standards Act implies the acceptance by Congress of the thesis that actual experience has proved it impossible to apply the assumptions of a free market mechanism to the labor market in a modern economy. The most significant result of the Act, therefore, is the protection it affords to the worker in bargaining with the employer.[9] By curtailing free competition in the labor market to a small degree, the Act helps labor to participate in developments in other markets that occurred long before. By fixing maximum hours and minimum wages, the Act assures the worker a living standard that, in numerous cases, he might otherwise be unable to secure, as a result of the intricacies of free competition on the supply side, and because of his weak position in the labor market.

Some employers and employer associations welcomed minimum wage regulation because it took the question of wages largely out of competition and saved them from the necessity of holding wages down to the level of irresponsible employers who are lacking in social consciousness.[10] In Great Britain it was found that some marginal producers were forced out of business when minimum wages were fixed by the Trade Boards during the past thirty-five years.[11] But, as a rule, they were replaced by more efficient units able to support the higher rates. Wages have risen; employment has not diminished. Similarly, a study made by the International Labor Office in 1924 showed that in Sweden, Switzerland, The Netherlands, Great Britain, France, the United States, and other countries, a reduction of hours stimulated improvements in equipment and, in general, organization of production and of work.[12]

Higher wages and shorter hours tend to influence the health conditions of the worker and his family favorably. Nutrition improves; sickness is less frequent. The experience of the past few decades proves that shorter hours have led to improvements in both the quantity and the quality of production, and that these improvements are due, at least in part, to the greater efficiency of the worker.

The Fair Labor Standards Act has an impact on the level of wages in the country as a whole, and on the redistribution of national income in favor of the worker at the expense of the employer. To what extent the distribution of national income will change, depends mainly on whether entrepreneurs find it possible to shift the increase in production costs, resulting from higher wages or shorter hours, to the consumer by increasing the price of the product. This presents much difficulty in the metal-mining industry, because gold and silver prices are fixed by law and copper, zinc, and lead prices are determined by world competitive conditions. On the other hand, part of the resulting decline in profits should be counterbalanced by an increase in management efficiency and worker productivity expected to result from the Act. Much will depend on conditions in individual industries and enterprises, but it can be assumed that some decline in profits will occur, despite the transfer of higher costs to prices, and hoped-for favorable effects on labor productivity.

B. THE VIEWPOINT OF THE MINING INDUSTRY

The non-ferrous metal mining industry, speaking through the pages of the Mining Congress Journal, official monthly publication of the American Mining Congress, was opposed to legislation of the type embodied in S. 2475 and H.R. 2700.

The first notice of such legislation, appearing on page 29 of the March, 1937, issue, states: "It is quite generally believed that the Administration Wage and Hour Bill will be aimed at industries charged with the 'sweating' of labor." The June, 1937, issue reports the introduction of the Black-Connery bills (S. 2475 and H.R. 2700), discusses their provisions, and predicts stubborn opposition. The July issue reports, in full, the statement made by Julian D. Conover, Secretary, American Mining Congress, before the Committee on Education and Labor, United States Senate, and the Committee of Labor, House of Representatives, giving the views of the non-ferrous metal-mining industry toward the passage of the Black-Connery Bill.

Mr. Conover said in part that the industry opposed the bill for the following reasons:

 1. They would impose upon the mining industry arbitrary legislation contrary to natural economic forces;
 2. Reduction of working hours, as contemplated in these bills, would seriously interfere with the operation of many mines;
 3. Imposition of onerous conditions through the operation of these bills would undoubtedly force cessation of production in many low-grade mines and marginal mining properties;
 4. Enactment [of these bills] would similarly tend to discourage prospecting and the development of new mineral deposits upon which the continuation of the mining industry depends;
 5. The bills would repose in a political board the power of life and death over American industries in all sections of the country; [and]
 6. Adequate enforcement of the provisions of these bills would, in our judgement, be impossible.

In the August, 1937, issue of the Mining Congress Journal, is it stated that "the Wage and Hour Board created by the bill is now limited to a minimum of 40 cents an hour in the fixation of wages and to a maximum of 40 hours per week. Also in Section 6 the exemption provisions have been rewritten and would now permit overtime employment if compensated for at the rate of one and one-half times the regular rates." The September issue said, "The Labor Standards Bill...has caused great concern to the mining industry in the past month."

Of serious concern to mining was the amendment introduced by Representative Ramspeck, which provided that between the hours of midnight and 6 a.m. persons employed shall be paid a rate of not less than one and one-half times the rate otherwise established by the board. As written, this provision would have interfered seriously with the long-established night-shift practice in the mining industry which is essential to economy of operation and the efficient use of production capacities. When the mineral producers were informed of the nature of the so-called "graveyard shift" amendment, they protested so vigorously that a number of representatives in Congress from western mining states and elsewhere held a special meeting and thereafter made a united protest to Chairman Mary Norton of the House Labor Committee. As a result, the Labor Committee, after having reported the bill to the House, met and agreed to accept a committee amendment from the floor striking out the "graveyard shift" feature.

In the October, 1937, issue of the Mining Congress Journal, one reads that the Senators and Congressmen from the South and the Southwest are almost a unit in determined opposition to the attempts of the administration counselors to inflict revolutionary wage-hour legislation upon the nation. "These counselors apparently fail to realize that industrial enterprises throughout the nation are widely different and that a blanket federal control of wages and hours is impractical, and difficult if not impossible of administration." The November issue reported that "The administration has stated that it believes further wage-and-hour regulation is essential to our present economic structure, in order that unemployment may be further decreased and a wider spread of purchasing power placed in the hands of the many. In view of the very definite opposition from the southern members of Congress, such a bill will need to be drawn in a moderate form...."

The December, 1937, issue (pp. 34, 35, 38) contains an article by Edward H. Snyder, a mine operator, entitled the "Analysis of Proposed Wage and Hour Legislation." He writes in part:

> After careful study my conclusions are that the real purpose of the authors of these bills [Senator Black and Representative Connery] was:
> 1. To centralize power to control all industry in a board to be appointed by the President.
> 2. To further promote the interests of those organizations that collect tribute from the working men of the United States.

The January, 1938, issue reviewed the numerous wage-and-hour bills and amendments that were offered to Congress and voted down. In February it was noted that additional wage-hour bills have been introduced but that action was postponed to allow a period for further investigation and crystallization of thought. It is noted further that since almost all organized groups, including labor, have shown dissatisfaction with the bill, Rules Committee Chairman John O'Connor's question, "Who wants the bill anyway?" still remained unanswered. The March issue said that "It now appears that wage-hour legislation activity will be deferred. It is widely recognized that because of geographical differences and varying industrial needs, the imposition of federal wage-hour control is impractical." The April and May issues pointed out that Representative Ramspeck has rewritten the bill but that "this controversial and unwanted measure" probably would not be enacted during that session of Congress. In the June, 1938, issue, the Mining Congress Journal reported that the Ramspeck Bill was voted down in the House Labor Committee, but that a new measure was reported which "is believed in some quarters unconstitutional" because no provision is made for review of orders on wage and hour standards.

In the July, 1938, issue of the Mining Congress Journal, however, an editorial, part of which follows, appeared:

The "Fair Labor Standards Act of 1938" on June 14 was approved by both Houses.

The law carries a 25 cents per hour minimum wage for the first year and a 30 cent minimum wage for the second year. Thereafter within seven years from the effective date of the Act the minimum wage increases to 40 cents per hour unless such a wage is found by the Administrator, in the Department of Labor, and his industrial boards to involve unemployment in an industry. The industrial boards are to be appointed by the Administrator with a representation of one-third each of management, labor, and consumers.

Of particular interest to the mining industries are the hours provisions...which are herewith reproduced:

"Maximum Hours

"Sec. 7. (a) No employer shall, except as otherwise provided in this section, employ any of his employees who are engaged in commerce or in the production of goods for commerce ----
  (1) for a week longer than 44 hours during the first year from the effective date of this section,

(2) for a work week longer than 42 hours during the second year from such date, or

(3) for a week longer than 40 hours after the expiration of the second year from such date,

unless such employee receives compensation for his employment in excess of the hours above specified at a rate not less than one and one-half times the regular rate at which he is employed."

C. CONCLUSIONS

The aim of Congress was to pass a law that would relieve unemployment and increase purchasing power. Unemployment may be alleviated by a reduction of the labor supply. The labor supply is reduced by (1) lowering the retirement age, (2) increasing the age of entrance of minors into the labor market, (3) keeping women out of industry, and (4) changing the character of unemployment by administration, as was done to a large extent in pre-World War II Germany.

Increasing purchasing power can be done (1) be increasing wage payments in private industry and (2) by public spending directed at maximum labor employment. Wage payments are increased either through increased wages for the same hours of work, or the same weekly wages for fewer hours of work.

The Fair Labor Standards Act raised the wages of the lowest paid workers, thereby raising the total purchasing power. The Act also lowered the legal hours of work in order to give employees more leisure hours in which to increase purchases of consumer goods. It made no attempt to regulate productivity of labor and efficiency in industry, leaving that to the collective bargaining agreements made between industry and organized labor.

The mining industry did not believe the Fair Labor Standards Act should be passed. The industry felt that it would be impracticable, or perhaps even impossible, to administer. The industry felt strongly that wages, working hours, and conditions in the mining industry were so far superior to those in the "sweated" industries that the proposed legislation should not apply to mining. The increased costs which would result from the Act must be absorbed largely by the producer, and any increase of costs in the mining industry results in a decrease of ore reserves, and of profits with which to search for additional ore reserves. Because an increase of costs results in a reduction of production, the mining industry felt certain that the Act would lead to addional unemployment and reduced purchasing power in its field.

Chapter III

EFFECTS OF THE ACT

The effects of the Fair Labor Standards Act of 1938 on the metal-mining industry can be charted through three different periods. The first period was while the mining industry and the Wage Hour Division were attempting to find out how the Act, as finally passed, applied to the industry; the second period began when enforcement of the Act was started; and the third period followed when the interpretation of the Act became clarified and enforcement difficulties diminished.

A. THE FIRST PERIOD

During the first period, the mining industry attempted to arrange its operations to comply with the requirements of the Fair Labor Standards Act. Because so many compliance questions were submitted by metal-mine operators to the Wage and Hour Division, an extensive unpublished study was made by the Research and Statistics Branch of that Division on the effects of the Act on the mining industry.

Information was collected by means of a questionnaire sent to all copper, zinc, lead, gold, and silver mines during the fall of 1939. In addition to the questionnaire, supplementary information was collected by two Wage and Hour Division staff members who visited most of the important mining districts of the United States, and who obtained interviews with operators, representatives of unions, individual employees, and government officials. The questionnaire was designed primarily to compare employment, payroll, and operating data for the first quarter of 1938 with that of the first quarter of 1939. These two periods, one before the passage of the Act, June 25, 1938, and the other after the effective date of the Act, October 24, 1938, were well chosen, for economic conditions were such that few other quarters have been so nearly comparable.

The information gathered is of great value, since answers to the questionnaire were received from 167 mines, employing almost half of the workers of metal mines and mills in the United States. The sample is not an accurate one because, while almost the entire mining industry of Arizona and Montana responded, only one-fifth of the industry in such important mining states as California, Nevada, New Mexico, and Utah sent back the answered questionnaire. However, the questionnaire shows, in general, how the mining industry had adjusted to the requirements of the Act.

1. Working Schedule Adjustments

The method of payment in metal mines always had been on a shift basis. In many mines, especially the smaller ones, the shift was measured by the completion of a cycle. The cycle in a mine includes the work of barring down loose rock to make the working place safe, loading and transporting the rock broken by the blasting at the end of the previous cycle, supporting the recently created opening (if necessary), drilling a round of enough properly placed holes to break the required amount of rock, loading the drilled holes (round) with explosive, and detonating the explosive. In some mines a miner was permitted to go home on completion of his cycle; in most of the others he was, and still is, permitted to loaf until blasting time after he has completed the earlier parts of the cycle. When the Act went into effect, many adjustments had to be made, since hourly wages and hours worked had to be calculated.

Mines[1] employing 21.6 percent of total employees in the sample in the first quarter of 1939 felt that they could afford to remain on their 1938 schedules and pay employees time and one-half for all hours worked beyond forty-four in any one week. On the other hand, mines with 9.2 percent of the sampled employees remained on their earlier schedules and disregarded the Fair Labor Standards Act entirely.

Another group of mines with 6.8 percent of the sampled employees made a so-called "bookkeeping adjustment" to the Act. The "bookkeeping adjustment" was an attempt to comply with the law, yet not increase or decrease the weekly wage paid a miner. Because miners' wages never had been calculated on an hourly basis, no hourly base existed. For the purpose of bookkeeping, then, the daily wage that the miner received was considered six hours straight time and two hours overtime. Therefore, a man could still continue working seven 6-hour days or forty-two hours straight time each week and fourteen hours overtime each week, or fifty-six hours per week as he had always worked. For instance, if the wage scale before the passage of the Act was $4.50 per day, the bookkeeping pay rate became 50 cents an hour for the first six hours and 75 cents an hour for the next two, or overtime hours. The miner was given the choice of working six hours for $3.00 or eight hours for $4.50.

The replies to the questionnaire show that many of the mines followed the spirit as well as the letter of the Act by requiring no overtime work from their employees. Only 16.6 percent of sampled employees worked forty-four hours a week or less in 1938, whereas 62.4 percent worked forty-four hours or less in 1939. The change toward the shorter work week was most noticeable among the larger mines, since one-third of the mines with one-half of the employees changed to a work week that required no overtime payments. Approximately one-third of the mines which reduced their hours began operating on a seven, seven and one-quarter, or seven and one-third hour six-day week; another third began operating on a five and one-half day week (two back-to-back weeks five and one-half days long were worked as eleven consecutive days each fortnight); and the remaining third changed directly to a five-day week.

Of especial interest is the number of western mines that set up "swing shifts" or staggered schedules. Four of the mines had been using swing shifts or staggered schedules in 1938 to maintain seven-day a week operation while working their employees six days per week, and they continued this schedule through the first quarter of 1939. Nine mines set up arrangements whereby the mine operated seven days of the week, while the men worked five or six days per week. Three additional mines kept in operation six days a week with the men on a five or five and one-half day basis. Consequently, 16 out of a total of 107 western metalliferous mines which answered the questionnaire used swing shifts or staggered schedules to maintain mine operations for a longer working week than that of the employees.

2. Cost Changes

What cost changes resulted from the adjustment of working schedules made by the various mines? The group of mines that chose to remain on their former schedules and pay the overtime penalty increased their costs in proportion to the penalty paid. Since four of the forty-eight hours in the first quarter of 1939 were penalty hours and had to be paid for at one and one-half times the regular rate, the mines that remained on a forty-eight hour week paid their miners for fifty hours each week. This amounted to a raise in labor costs of 4 percent. The mines that remained on a fifty-two hour week paid their miners for fifty-six hours each week, with an increase of labor cost of 8 percent. The mines that remained on a fifty-six hour week increased their labor costs 11 percent. Since labor costs average over half of the total cost of mine production, a rise in labor cost of 10 percent is equivalent to an increase of over-all costs of 5 percent.

The mines that disregarded the Act were not affected immediately by it. The mines that chose to adjust to the Act by keeping more detailed pay records were affected by the Act to the extent of the additional bookkeeping cost.

The greatest adjustment of working schedules was made by the group of mines that shortened their work week to or below the unpenalized maximum required by the Act. Some mines that shortened their work week did it by shortening their working day, though most reduced the days worked. The decision was a difficult one to make, since the shortening of the work day imperiled the completion of the cycle each day. The cycle includes all the operations necessary to blast a round of holes, and the blasting is the last part of the cycle. In most underground operations, it is permissible to blast only at the end of the shift. If the cycle is not completed in one shift, it cannot be completed until the end of the second shift.

The mines that reduced the number of days worked were faced with an increase of service and overhead costs. The questionnaire showed that almost all underground mines that worked fewer than seven days a week carried on some service operations during the off period. The most common service operations include pumping the water from the mine, ventilation of the mine to keep it free from harmful gases, repair and replacement of timbers against continuous rock pressures, and patrol of the mine in search of fires, fire hazards, broken air, water, and power lines, and clogged drainage ditches. If the mine is a shaft mine, the hoist must be in operation at all times to provide access for those men who work underground on non-operating days.

Overhead expenses are almost constant no matter whether five or seven days are worked per week. They include executive, administrative, engineering, and office expense; interest and depreciation; and other costs, such as taxes on ore reserves.

The service and overhead costs that continue without regard to production commonly range from 10 to 30 percent of the total cost. Where production was decreased one-sixth by dropping from a six-day week to a five-day week, total cost per ton was increased 2 to 5 percent from this cause. Only if the productivity of the worker increases by 2 to 5 percent will the total cost of production remain unchanged.

The questionnaire attempted to determine whether overhead costs and productivity did increase for those mines which reduced the work week of their employees. Apparently all mines that decreased their work week experienced increases in overhead. The mines that reported a decrease in overhead cost were those that continued operating as many days per week as before the inception of the Act, but reduced the work week of their employees. The questionnaire showed further that the productivity of the employees rose 9.7 percent and that the average labor cost dropped 2.9 percent. The Wage and Hour Division report comments, "It is not possible to assign definitely causes for increased productivity. Several mines report that it was due to a reduction in development work and hence a greater proportion of employees were engaged in productive work."

The mine operators did not agree with the conclusion of the Wage and Hour Division that productivity of labor, because of a decrease in labor's working hours per week, increased enough to offset the increase of costs caused by the Act. What follows is based on correspondence with many mine operators concerning this subject, opportunities to discuss the subject personally with them, and study of the pages of the mining journals for their published comments. The comments of the mine operators were all alike in agreeing that costs rose. Some believed that productivity increased, but its cause was not the reduction in the work week. Technological improvements continued to be made and operated to improve labor productivity. Other factors, such as decreased employment opportunities, increased labor productivity during the first quarter of 1939. But few operators believed that the Act had any immediate effects on the productivity of labor.

The study of the effects of the Act on the first period, during which the mining industry attempted to comply with the Act, showed that the Act had generally raised costs. The increase of costs ranged from nothing among the 16 percent of the mines which worked five days or fewer before the Act was passed, to a maximum of 6 percent among those mines that worked seven days a week and paid the overtime penalty. The mines that conformed to the spirit of the Act and paid no overtime found that costs rose from 2 to 5 percent because of an increase in service and overhead costs per unit of production.

3. The Administration of the Act

As already stated, the non-ferrous metal mining industry, as a whole, attempted to comply with the most understandable interpretation of the Act that they could obtain. Compliance with the Act was a practical problem, and the methods for compliance by the mining industry were not specified in the Act. The mining industry, like many other industries, flooded the Wage and Hour Division with requests for information, stating typical operational problems and requesting authoritative rulings as to the applicability of the Act. However, Congress had given the administrator of the Act no power to make rulings. Every operation of enforcement must be litigated in the courts.

Under the Act the administrator was given the power to prescribe certain regulations, and to hear and to investigate complaints. The regulations prescribed by the administrator may be revised after a hearing has been held with all parties concerned. The regulations concerning the mining industry that were set up include: records to be kept by employers; definition of industries of a seasonal nature; and definition of reasonable costs of board, lodging, and other facilities furnished as a part of wages. The Wage and Hour Division prepared and distributed complaint forms, with the assurance that the name of the complainant and his complaint would be kept confidential. The actual responsibility for the complete enforcement of the Act rested with the employees and the labor unions. However, both the CIO and the AFL asked their members to proceed to exercise their rights under the Act with reserve until the constitutionality of the law had been established in a suit brought by the government.

To assist harried employers in complying with the Act, the Wage and Hour Division hit upon the expedient of issuing interpretative bulletins. In addition, frequent press releases covering court decisions, new interpretations of the Act, and clarification of the administrative policy were circulated. A cautious policy, which included numerous hearings before a definite rule was set up or an interpretation was released, assured realistic administration of the Act.

B. THE SECOND PERIOD

The second period is that in which the applicability of the Act became apparent and enforcement began. At first, the mining industry, along with other seriously affected industries, attempted to avoid the law or have it amended rather than to violate it. Many amendments to the Act were proposed but only two, neither seriously influencing the mining industry, were passed. Other methods by which the industry attempted to obtain relief were by obtaining rulings through court decisions or interpretations from the Administrator.

One hope of a major portion of the mining industry was that it would be declared by the courts an industry in intrastate commerce, and therefore not affected by the Act. In a suit brought by the National Labor Relations Board, the Idaho-Maryland Mines Corporation, a gold-mining company of California, had been ruled in intrastate commerce in a decision by the Federal Court of Appeals. Thereafter all mines that shipped their product to a smelter within the same state, and did practically all their buying within that state, hoped to be freed of the burden of the law. The administrator of the Wage and Hour Division, however, warned the mining industry that he would attempt to have the Act enforced against all mines despite the decision in the Idaho-Maryland case. Most mines, including the Idaho-Maryland, decided that the potential penalties of back pay were so great that they could not afford to ignore the Act.

The administrator's warning was substantiated by a unanimous decision of the Supreme Court which upheld the constitutionality of the Fair Labor Standards Act of 1938 in the case of the United States v. Darby, 312 U. S. 100. The Court was composed of Charles Evans Hughes, Chief Justice, and James Clark McReynolds, Harlan Fiske Stone, Owen J. Roberts, Hugo L. Black, Stanley Reed, Felix Frankfurter, William O. Douglas, and Frank Murphy as Associate Justices.

The Justices held, among other things, that the minimum wages and maximum hours provisions of the Act were within the power and consistent with the Fifth and Tenth Amendments; that, while manufacture is not in itself interstate commerce, the shipment of manufactured goods interstate is such commerce and the prohibitions of such shipments by Congress is a regulation of interstate commerce; that Hammer v. Dagenhart, 247 U.S. 251, in which a child labor law was held unconstitutional was overruled; and that the wage and hour provisions of the Act do not violate the due process clause of the Fifth Amendment.

Comments by the legal profession upon the decision, especially upon that portion that specifically overrules Hammer v. Dagenhart, generally approved the "enlarged scope of the commerce power" given to Congress.[2] The legal profession now believes that even the gold mines will be held to be engaged in interstate commerce, no matter to what state the gold is shipped after it is mined or where the mine buys its supplies and equipment.

That portion of the mining industry which chose to make "bookkeeping adjustments" to comply with the law, later learned that it did not meet the Administrator's approval. Philip B. Fleming, Administrator, Wage and Hour Division, wrote in the December, 1940, issue of the Mining Congress Journal as follows:

> The Act is not intended to force employers to pay more money to their employees who work overtime. The Act seeks to induce employers to reduce hours by making overtime hours more costly. It was the intention of Congress, by putting a penalty upon overtime work, to encourage the employment of additional labor. It was a spread-the-work device. It was assumed that imposition of the overtime penalty would encourage employers to eliminate overtime and employ extra workers. The employer does not comply with section 7 of the Act unless he pays his employees time and one-half for each hour they work in excess of 42 hours a week.

In this opinion the Administrator was guided by the wording of Section 18 which states that "No provision of this Act shall justify any employer in reducing a wage paid by him which is in excess of the applicable minimum wage under this Act...."

The mining industry was sufficiently impressed by the Administrator's opinion so that the "bookkeeping adjustment" method was abandoned. Nevertheless, the Supreme Court decided on June 8, 1942, in a 5-4 decision, that it was legal for a corporation to negotiate an individual contract with each employee guaranteeing him a basic hourly rate of pay, and also guaranteeing him a certain weekly rate of pay that is larger than the weekly pay normally would be even with some time and one-half for overtime (316 U. S. 624, Walling, Administrator of the Wage and Hour Division v. A. H. Belo Corporation). In many respects the arrangement that was ruled legal by the Supreme Court resembled the "bookkeeping adjustment."

At one isolated Oregon mine an attempt was made to continue working seven days a week under the Act without changing the working days per week or wages of the employees. The employees were told that if they could produce a satisfactory tonnage at the face in six hours instead of eight hours that no changes in pay would be made. The Wage and Hour Division, on a complaint, investigated to determine the validity of the practice under the Act. The investigation showed that the miners entered the mine either at 7:00 a.m. and came out at 3:30 p.m., or entered at 7:00 p.m. and came out again at 3:30 a.m. Eight and one-half hours were spend underground, of which thirty minutes was a lunch period, leaving a net of eight hours per shift. But the employees only showed six hours on their time slips, which they kept themselves and turned in each day. The company contended that this manner of time keeping was proper, since approximately two hours were consumed in travel time from the entrance of the mine to the working place and return. The contention of the Wage and Hour Division was that the entire time consumed from portal to portal was time spent at work, and that the company had failed to pay extra for overtime and had failed to keep on its records the time actually worked. The Division was upheld by the United States District Court for the Western District of Washington, Northern Division, in a judgment entered for the plaintiff in Fleming v. Cornucopia Gold Mines, May 3, 1940.

In later discussions the terms "face-to-face," "portal-to-portal," and "collar-to-collar" frequently are used. Face refers to working face or working place in the mine. Portal refers to the entrance of a mine when that mine is entered by a horizontal opening. Collar refers to the entrance of the mine when the mine is entered by an inclined or a vertical opening.

1. Work Defined

Rulings, such as made in the Cornucopia decision, led to the broader question, what is work? Section 7a(3) of the Act requires that the employer shall not employ such employees "for a work week longer than forty hours...unless such employee receives compensation for his employment in excess of the hours above specified at a rate of not less than one and one-half times the regular rate at which he is employed." According to section 3g, "Employ includes to suffer or permit to work."

Let us examine how the state courts have felt toward travel time as time worked. The earliest one, Ex Parte Martin, 157 California 59, 106 p. 235 (1909), a decision of the Supreme Court of California, is the single decision holding that underground travel is not part of the workday. Since then the Montana Eight Hour Law was construed to include "portal to portal" principle in the case of Butte Miners Un-

No. 1 v. Anaconda Copper Mining Company, 112 Montana 418, 118 p. (2d) 148 (1941).

That the "portal-to-portal" basis had been applied to coal mines, prior to the current war, in Austria, Belgium, British Columbia, Canada, Czechoslovakia, France, and the Dombrava and Cracow areas of Poland is reported in "The World Coal Mining Industry."[3] According to the same reports the Netherlands, the Upper Silesian Region of Poland, Alberta, Canada, and Spain included either the descent or the ascent in the computation of underground hours worked. In Great Britain, the shift begins when the last worker of the collective working group enters the cage to descend and ends when the first worker leaves the cage on returning to the surface.[4] So a considerable part of the travel time is included in the hours worked since the group would naturally tend to enter together and leave the mine together at the end of the shift. The Hours of Work (Coal Mines) Convention of 1935 adopted by International Labour Code (1939), pp. 81-83, provides in Article 3 that:

> Hours of work in underground hard coal mines means the time spent in the mine, calculated as follows - (a) time spent in an underground mine means the period between the time when the worker enters the cage in order to descend and the time when he leaves the cage after reascending; (b) in mines where access is by an adit the time spent in the mine means the period between the time when the worker passes through the entrance of the adit and the time of his return to the surface.

This Convention was ratified by Cuba and Mexico, both of which together do not produce one-tenth of one percent of the world's annual coal production. The Convention does not go into force until six months after the registration or ratification by at least two of the following important coal producing countries: Belgium, Czechoslovakia, France, Germany, Great Britain, The Netherlands, and Poland.

In the United States after the passage of the Fair Labor Standards Act, the Administrator of the Act was under great pressure from the mining industry and other industries to define what Section 39, "to suffer or permit to work" meant. In Interpretative Bulletin No. 13 of Wage and Hour Division dated November, 1940, the Administrator indicated the course which he intended to follow with respect to the determination of employee's hours of work unless authoritative rulings of the courts directed otherwise. These interpretative bulletins had been carefully prepared with legal aid and therefore have been accorded especial weight by the courts.

Interpretative Bulletin No. 13 stated that travel time will be considered work if it can reasonably be described as "all in a day's work." If a crew of workers is required to report for work at a designated place at a specific hour and all the employees are then driven to the place where they are to perform work, the time spent in riding to such place and return should be considered a day's work. Since Interpretative Bulletin No. 13 was so general in scope and the practices of the metal mining industry were so unusual, a great demand arose for the application of the general principles to the actual situations as found in underground metal mining. The Wage and Hour Division therefore conducted a thorough investigation, including field surveys and questionnaires, reports by the Division's Regional Directors, and public conferences at Salt Lake City, December 11 and 12, 1940, and Birmingham, January 14 and 15, 1941. Many employers, mining associations, and labor unions filed supplementary statements and briefs with the Division.

The investigation resulted in a "Report on 'Hours Worked' in Underground Metal Mining," prepared by Harold Stein and Rufus G. Poole, and released as G-133, March 23, 1941, by the Wage and Hour Division of the United States Department of Labor. The report showed how the metal mining industry defined, in a practical way, hours worked:

> The miner begins his day by arriving on company property, goes first to the change house where he removes his street clothes and puts on working clothes. The use of the change house (which is almost universal in metal mining) by the miner is optional and with rare exceptions free. The time spent in the change house is variable but ten or fifteen minutes would seem to be not uncommon.
>
> After changing his clothes the miner ordinarily walks to the portal or collar of the mine to check in (he reports his presence and receives a tag or check of metal or other material which he keeps on his person until he checks out at the end of his shift), stopping on the way at the tool house to pick up tools and other small supplies and at the lamp house to get his electric lamp or to buy carbide. [In the writers' experience carbide was always supplied free.] In some mines all tools and in many mines all large tools are taken into the mine for the miner as a matter of safety.

Usually the lamp house and the tool house are located close either to the change house or to the portal. Thus the additional time consumed by the miner in obtaining tools or carbide or lamps is only two or three minutes on the average beyond what it would otherwise take him to go from the change house to the portal.

The descent into a mine, if vertical, is by a cage. If the adit is inclined, the descent is made by a train, or skip, or on foot. Descent by conveyance is normal in large mines, and almost universal in deep mines. Descent on foot is common in small and shallow mines reached by an easy incline.

There is normally some wait for the cage or man-trip depending on the number of men and the available transportation facilities. This wait may be as much as twenty or thirty minutes. The descent seldom takes more than five or ten minutes. The miners proceed without delay to their working places, stopping only to pick up tools and supplies where necessary. The time required to reach the working point for the actual miner varies with the distance from the station to the working face, the method of transportation and the difficulties of the terrain. The actual elapsed time from arrival at the first station in the mine to the arrival at the working face thus will vary in different mines and for different miners in the same mine from five minutes or less to an hour or more. An average travel time of twenty or thirty minutes is common.

It is seen then that the underground metal miner, especially in mines that are deep and have extensive workings, spends considerable time in reaching his working place. Because of the well-established custom that when a miner, or mining crew, completed their cycle they had "put in a shift," little attention had been paid by the metal mining industry to the exact number of hours in a day or what exact operations constituted work.

On the other hand the large and powerfully united group of coal miners always had been explicit in their collective bargaining agreements that work began at the face and ended at the face. The "face to face" method of calculation of time worked had been customary only in those metal mines that were in coal mining regions, such as the iron mines of Alabama. Of 115 metal mines, with 22,300 employees reporting, only 4 with a total of 74 employees operated on a "face to face" basis.[5]

Undoubtedly the "face-to-face" method of computation of working time is historically the oldest, since all mines originally started at the surface and, for at least some period, the time spent in travel from the collar to the face and return was negligible. As the mines became deeper the metal miner's day in western United States was considered to begin at the time he entered the mine, and to end when he blasted the round and left the face. The miner was not paid for the time spent in returning to the surface. This interpretation of the length of the shift was probably in most general use in underground metal mines in the United States at the time of the passage of the Act.

Some mines employed a variation of this system and considered the start of the miner's day at the time he reached the working place and the end when he again reached the surface. This interpretation was required by law in the state of Nevada. However, during three years spent in working in and visiting underground mines in Nevada one of us seldom saw that method in force. A much more common method of calculation of working time, and one that was employed by the larger operations, was to consider that a man's shift began at the portal or collar and ended when he returned to the underground station before being transported to the surface. In none of these systems of payment was the time spent eating lunch, usually one-half hour, considered time worked.

In two very important non-ferrous metal mining states the miners are required by law to be paid on a "collar-to-collar" basis. Both the Arizona and Utah statutes require that underground miners should not spend more than eight hours underground in any one day, except in case of an emergency, and that these eight hours should be measured from the time a man enters the mine until the time he again leaves the mine.[6] In such a case the miners eat lunch on the company's time.

To be more specific, men are usually paid from the time the whistle blows in the morning which is generally coincidental with the lowering of the first cage-load of men into the shaft. In mines that have large work forces and inadequate transportation facilities the men who ride the second, third, or fourth cage into the mine actually spend as much as twenty minutes of their working time waiting on the surface before being lowered into the mines. Also, those men who come up in the first cage at the end of the shift may arrive on the surface before the official quitting time. It readily can be seen that precise definition of time of work in non-ferrous metal mines was difficult. A definite trend towards a shorter work day had been in evidence, as well as the inclusion of more of the parts of the day as time spent at work. At least one union contract specified that the miner be paid for all time that elapsed between the time he left the change house until he again returned to the change house at the end of the working day.

Many employers contended that miners should be paid only for time spent at the working face, for that is the only part of the day from which the employer derives any benefit. They also stated that it is impossible to shorten the time allotted to the mining cycle without seriously disrupting the operations of the mine, reducing profits, and generally interfering with the healthy development of the industry, including the welfare of its employees. In support of their contention they cited the language of the Fair Labor Standards Act, particularly the definition of the word "produced" in Section 3 (j) as follows:

> Produced means produced, manufactured, mined, handled, or in any manner worked on in any state; and for the purposes of this Act an employee shall be deemed to have been engaged in the production of goods if such employee was employed in producing, manufacturing, mining, handling, transporting, or in any other manner working on such goods, or in any process or occupation necessary to the production thereof, in any state.

They asserted that this definition indicates that only productive work should be counted as "hours worked" under the Act. They also cited what they consider both the dictionary and the ordinary application of the term "hours worked" as indicating that "hours worked" should apply only to the so-called production hours at the face. These arguments overlook the fact that the phrase "hours worked" does not appear in the Act. The question before the Wage and Hour Division was, "When is the miner acting in the capacity of employee?"

In the "Report on 'Hours Worked' in Underground Metal Mining" previously referred to, the discussion continued:

> It is not necessary to dwell on the general economic considerations. Obviously Congress realized in passing the law that compliance with its provisions would require certain adjustments. Further, the assertions made about loss in production occasioned by a "collar-to-collar" day are based on the assumption that all other factors affect-

ing production will remain constant; however, it can be assumed that insofar as an adjustment may be necessary, all other factors will not remain constant as the mine operators allege. Indeed all other factors ordinarily do not remain constant when there is a change in one of the major factors involved in a total operation. An interesting and directly relevant illustration of this is to be found in the statement submitted by the Phelps-Dodge Corporation with reference to its United Verde Branch which operated a large mine in Arizona. The question asked in the questionnaire was the following:

> "Total time spent at working face and an estimate of the minimum time required for completing a round of work."

In reply the General Manager of the corporation submitted the following:

> "Average time spent at working face is 6 hours, 30 minutes. Minimum time required for completing round of work cannot be answered, as rounds of work are adjusted in accordance with the time available for performing them."

While the adjustment referred to will be more difficult in some cases than in others, it is the type of adjustment that normally and necessarily results from the enactment of any new piece of social legislation. Finally, it must be recalled that thousands of miners are employed by mine operators on the very basis which is described by other mine operators as prohibitive.

It is also not necessary to consider in detail the precedents established by state 8-hour laws and decisions thereunder. No single one of the state statutes is worded like the federal statute nor does any have precisely the same purpose and intention. These laws limit the underground work day by a flat limitation; the Fair Labor Standards Act tends to limit the total work week by an overtime penalty. The state laws reflect the general opinion that a limitation of underground time is desirable; but they cannot be accepted as controlling on the particular points at issue here.

The basic decision to be made is whether in the light of the policy of the Fair Labor Standards Act, working time for the purpose of section 6 and section 7 is to be construed as merely the time in which a particular set of activities are performed or whether it should also include the time which the employee gives to his employer and in which his activities are prescribed by his employer's rather than his own needs and desires.

The contentions of the unions cannot be accepted in certain respects. Travel time whether on foot or by cage or train is time devoted by the employee to the service of his employer; it is not subject to the employee's discretion or control and it is not for the employee's benefit. The contention that the inclusion of travel time as part of the work day or work week logically requires the inclusion of travel time from the employee's home to the mine does not merit serious consideration. Generally speaking, the employee has choice as to where he lives and as to his method and time of reaching the mine.

This finding that the miner's day begins when he reports at the collar of the mine is contrary to the claims advanced by two unions [The International Union of Mine, Mill, and Smelter Workers, affiliated with the C.I.O., and the Federal Labor Unions, affiliated with the A.F. of L.]. From all the evidence it appears that the bathhouse is a convenience furnished by the employer, usually free, for the benefit of the employee. In many mines, furthermore, there is no bathhouse at all, and in mines where there is a bathhouse, its use is optional. In view of these facts there appears to be no justification for finding that the employer should pay for time spent in the bathhouse. Similarly the time spent in walking from the bathhouse to the mine portal is consequent upon the employee's beneficial use of the bathhouse itself. So far as the employer is concerned the employee could report directly to the mine portal and some employees ordinarily do so. Under these circumstances it is not reasonable to accept the claim of the A.F. of L. that working time begins at the moment the miner enters the bathhouse; nor is it reasonable to accept the contention of the C.I.O. that the day begins when the miner leaves the bathhouse.

Both unions contend that the work day should include the lunch period. The mine operators contend that the lunch period should not be counted as part of the work day. If in any particular factual situation the lunch period is used for the benefit of the employer, as in getting tools, there is no lunch period and there is no problem involved. More basic considerations urged by the unions are that the miner is subject to the hazards of his occupation and to the disadvantages and inconveniences inherent in being underground during the lunch period as well as during the rest of his time underground. There is much weight in these contentions. Even where lunchrooms are provided underground the difference between such a lunchroom and an ordinary restaurant on the surface are great indeed. Some of the hazards of underground employment continue and it is unquestionably less relaxation for a miner to eat underground than it would be to eat on the surface. His movements are necessarily limited and his opportunity for rest is constricted. However, these are matters of degree. In spite of these disadvantages there remains a sharp distinction on the one hand between the time when the miner is drilling, etc., or traveling to his place of work, and on the other hand, the lunch period. There is a considerable element of relaxation involved and except for safety restrictions (which are also imposed in the bathhouse) the miner is free from supervision for the lunch period.

Summary

The workday in underground metal mining starts when the miner reports for duty as required at or near the collar of the mine, and ends when he reaches the collar at the end of the shift.

The workday also includes the aggregate of time spent on the surface in obtaining and returning lamps, carbide and tools, and in checking in and out.

The workday does not include any fixed lunch period of one-half hour or more during which the miner is relieved of all duties, even though the lunch period is spent underground.

The report and conclusions reached in the summary were approved by Philip B. Fleming, Administrator of the Fair Labor Standards Act, May 1, 1941:

The Supreme Court, on March 27, 1944, upheld the opinion of the Administrator in the case of Muscoda Local No. 123, Etc., Et.Al. v. Tennessee Coal, Iron and Railroad Company; Sloss Red Ore Local No. 109, Etc., Et.Al. v. Sloss-Sheffield Steel and Iron Company; and Reimund Local No. 121, Etc., Et.Al. v. Republic Steel Corporation. The Court ruled, in a 7-2 decision, that the Fair Labor Standards Act requires "portal-to-portal" pay for iron ore miners and the wording of the majority opinion, written by Justice Frank Murphy, left little doubt that the same decision would be applied in any similar cases concerning miners.

2. The Effect of the Second Period

The second period was that in which the applicability of the Act became apparent and enforcement began. It was established that the entire mining industry was engaged in interstate commerce, and therefore subjected to the requirements of the Act. Even more important to the mining industry was that the Act could not be interpreted without a definition of the term "hours worked," and when the courts defined "hours worked," they defined them as "collar-to-collar" or "portal-to-portal" hours.

Much of the non-ferrous metal mining industry prior to the passage of the Act had not worked on the "collar-to-collar" principle. That portion of the industry, then, experienced a serious rise in costs because of the ruling that work started and ended at the entrance to the mine. Because of the reduction in the miners' working hours without reduction in pay, the increase in labor cost rose from 2 to 25 percent.

The effects of the first period of the Act were shown to increase total mining costs up to 6 percent. The effects of the second period of the Act increased over-all costs up to 12 percent. Consequently, the direct effects of increased wage rates because of the overtime penalty, or increased overhead cost because of the reduction of operating days, added to the indirect effects of shortening the miners effective work day, led to an increase of mining costs of 15 percent in areas where metal mining had been carried on in a traditional manner before the Act was passed.

C. THE THIRD PERIOD

The third period was that in which the interpretation of the Act became clear and enforcement difficulties diminished. "Hours worked" had been defined and the "portal-to-portal" principle accepted as binding for all mines. No further discussion arose about the payment of time and one-half for work beyond 40 hours a week. Schedules were adopted by all mines to fit the new regime. But had wage and hour questions become static again?

Edward H. Snyder, in 1937, stated (p. 20) that the real purpose of the Act was "to further promote the interests of those organizations that collect tribute from the working man of the United States." In 1939, Otto Nathan wrote that the most significant result of the Act is "the protection it affords to the worker in bargaining with the employer." It has been shown that the cost of operation in the mining industry

EFFECTS OF THE ACT

has been increased considerably by the Act. Yet, Mr. Nathan had predicted that the most significant result of the Act would be the increased bargaining power that it afforded the employee.

Many letters and personal contacts with operators in the mining industry confirm both Snyder's and Nathan's predictions. Unionism was not important among the metal miners before the passage of the Act. After the passage of the Act, the unions were able to organize the employees of most of the mining companies by asking that the miners receive the same "take-home" pay for the 5-day week that they had received for working the longer week. The unions also promised that they would make complaints and insist on action in the many cases where the operators were not abiding strictly by the Act. In most of the Act violation cases that went before the courts, the employees, generally represented by their unions, were victorious. The unions gained great prestige and enthusiastic membership support among the miners.

With their new-found powers the employees, through their unions, bargained with the operators and were able to obtain many privileges in addition to those granted in the Act. Double time for Sunday and holiday work was written into many contracts and, then, during the war, double pay for the seventh day worked each week was made mandatory by President Roosevelt for all employees engaged in the production of war goods. Few mines gave their employees vacations with pay before the Act was passed; now, most union contracts with mine operators require that employees receive vacations with pay each year. Many contracts provide that miners be paid for their lunch period, and that eight hours "portal-to-portal" means that the miners shall be underground a total of no longer than eight hours, including the time they take for lunch. Additional pay for the portion of the shift worked between 6:00 p.m. and 6:00 a.m. has been granted most mine employees through collective bargaining. Many mines supply their employees with transportation to and from the mine and the nearest town. These, and many more, advantages have been gained for the employees through collective bargaining in addition to the numerous pay raises that they have received. In a letter, the International Union of Mine, Mill, and Smelter Workers informed us in part that, "Although the 40-hour week has of course had some effect on the miner's weekly earnings, a much more decisive factor has been the increase in hourly wages won through collective bargaining."

The third period of the Act, then, was the period in which

production costs rose the most. Only part of this large rise in costs can be attributed directly to the Act, because many other important forces at work during the past decade also affected hours and wages. The rights of employers and employees under the Act have been clearly established and, while no additional effects of the Act itself can be expected, its indirect effects which resulted in a reduction of hours and an increase in wages may be expected to continue.

D. SUMMARY

The Fair Labor Standards Act of 1938 resulted in increased costs for mining. The Administrators of the Act were frank in acknowledging that mining costs have risen. They said that any social legislation such as the Act was bound to cost industry money, but that it was justified because it improved the welfare of a large group of people.

The increases in cost that are directly attributable to the Act are: (1) The increase in wage payments incurred by the mines that continued working beyond forty hours each week and that paid the overtime penalty; (2) the increase in unit service and overhead costs incurred by the mines that reduced their work week in order to keep from paying the overtime penalty; (3) the increased costs that resulted from the shorter work day that was granted to miners because of the definition of the term "hours worked."

All three of these cost-increasing factors affect the mining industry more seriously than they affect most other industries. Because of their situation many mines were not in a position to shorten their work week. If a shortened work week had been instituted by mines far from large towns, many of the miners would have gone to more populated places on their days off, never to return to work at those mines. At many mines which attempted to decrease their work week and keep up production by working a larger crew during the shorter week, the cost of the additional housing, additional equipment, and enlarged mine installations would have been greater than the cost of the overtime penalty. The mines that were able to shorten their work week had to continue paying the large service and overhead charges which are unique to the mining industry.

The mines that had not worked on a "portal-to-portal" basis before the Act were similarly penalized for being mines rather than other types of business. With few exceptions, the mining industry is the one whose employees must spend the most unproductive time traveling to their work on the company's time. Since the mining industry "really is different" in many respects from other industries, the direct cost increases caused by the Act were serious.

Cost increases indirectly attributable to the Fair Labor Standards Act are those that came about through increased unionism and collective bargaining of the employees with the employers. These cost increases, in most instances, are greater than all the direct cost increases combined. But only a portion of them can be charged to the Act and the remainder must be charged to other parallel-acting forces.

What effects do increased costs have on the size and profitability of the mining industry? The producers of copper, lead, zinc, silver, and gold do not control the selling price of their products. Gold and silver prices are controlled by governmental fiat; copper, lead, and zinc prices are controlled by world-wide supply and demand. The supply consists of foreign metal as well as domestic metal, and also all other competitive metals and competitive substitute products; e.g., aluminum for copper in transmission lines, aluminum and tin for lead in toothpaste tubes and foil, plastics for zinc in decorative automobile parts.

If production costs per ton of ore increase and the selling price per pound of metal does not increase, then the mine operator must produce ore of a higher metal content to continue to operate at a profit. This can generally be done in any mine since most mines have various grades of ore. At constant metal prices a mine may have ten million tons that can be mined for $4.00 a ton and five million that can be mined at $5.00 a ton. Consequently an increase in over-all costs of 25.0 percent from $4.00 to $5.00 per ton may reduce the ore reserves 50 percent or from ten million to five million tons.

The inevitable tendency of any increase in metal mining costs is to reduce the tonnage of ore reserves available for profitable exploitation. When ore reserves are reduced, either the size of the operation is reduced or the length of time until the orebody is exhausted is shortened. Both of these effects tend again to increase costs.

In a highly profitable operation, which is producing from a super-marginal orebody, the chief effect of an increase in costs is a decrease in profitability. Because the profitable companies generally are the ones with enough faith in the future of the industry, and with sufficient resources, to search for additional orebodies, a decrease of profits will result in a decrease of prospecting and, in the long run, fewer new orebodies discovered.

An increase in the unit cost of mine production, then, tends to reduce the size of the industry, the volume of reserves, and the number of employees at work. Therefore, the Fair Labor Standards Act of 1938 tends to decrease the importance of the non-ferrous metal mining industry in the United States, because of the increases in costs attributable to it.

Chapter IV

INVESTIGATION OF METHODS FOR SURMOUNTING INCREASED COSTS

The previous chapter has shown that the Fair Labor Standards Act of 1938 resulted in a substantial increase in costs for underground metal mining. In this chapter possible methods of overcoming the increase of costs will be investigated.

Decreases in unit costs can be brought about in two ways:

1. By an increase in the productivity of labor per man-hour of productive time at work, and
2. By an increase in the proportion of productive time to non-productive time while the employee is at work.

Considerable attention has been directed in the past to the improvement of labor productivity; but not enough heed has been paid in metal mines to the problem of a more complete use of the time for which labor is paid.

A. IMPROVEMENT OF LABOR PRODUCTIVITY

Labor productivity is increased by the application of technological changes. A number of types of industrial technological changes are recognized, most of which may have application to the mining industry.[1]

One change is the installation of new types of equipment designed by specialized machinery manufacturers. In the mineral industries this type of change is one of the most important. A number of mining machinery manufacturers continuously improve old types of machines or build new types which they hope will find a ready market because of their increased efficiencies. The mining industry cooperates with these manufacturers in suggesting improvements and permitting trials of the improved product.

Another technological change is the installation of machinery designed by the user. This, again, is important in mining since most of the large mines have well-equipped shops, and sometimes foundries, which are used not only for the repair and maintenance of existing equipment but also for the development of new types of machines and improvements on the old.

A third type of change is the operation of old machinery at higher speeds or in new combinations. Some few of the machines used underground have variable speeds; this kind of technological change is not readily adaptable to mines. But in some cases, haulage or transportation equipment might be speeded up, or operated in new combinations, with increased efficiency.

The fourth type, the introduction of new tools and fixtures, is again important in underground metal mines, because mining operations have so many individual differences that "gadgets" often tend to increase output. An example of such an improvement, recently noticed, is the use of a simple trough full of waste oil through which the scraper cables constantly run, with the result that cable life has been increased considerably. Miners are notorious gadgeteers.

A fifth type is the use of new or improved materials; for instance, the adoption of aluminum or high-strength alloy steels in order to reduce the dead load and increase the pay load of mine-hoisting equipment.

A sixth type of technological change is the improvement of manual methods. This includes time and motion studies of workers while at work. Very little of this has been done in underground mines; therein lies a fertile field for future improvement. Where this change includes the improvement of illumination (the increase in visibility of objects), and the reduction of dust, high temperature, and accident hazards, great strides have been made in underground mines. However, only in South America have the writers seen the use of scheduled mid-morning and mid-afternoon rest periods in order to ward off fatigue of the miner.

The seventh type is the subdivision of complex tasks. In most mines one crew performs all the operations required for breaking, loading, and transportation. With increased mechanization both loading and transportation duties have been taken, in many cases, from the man who does the breaking.

An eighth type of technological change is the mechanization of manual work. While this has reached a high stage of efficiency in manufacturing plants, and even in mills and open-pit mines, much heavy labor still is done by hand in mines. At least part of the reason for this is that the barring down of loose rock can never be a mechanical job, and the standing of a set of timbers in its proper position must always be done manually.

The ninth and last type is industrial research. The growth in its use is proof that it is a technological change which results in the reduction of costs in industry. However, it is a change which is as yet being applied by only a few of the mining companies. Mining had been quite profitable in the nineteen twenties; consequently, little research seemed necessary. In the thirties, because of lack of profits, there were no funds to support research.

All these types of technological changes will tend to be increasingly employed by underground mine operations in order to overcome the increased costs attributable to the Fair Labor Standards Act. Only in a few mines are conditions such that some form of technological improvement cannot be applied successfully. The large majority of mines will continue to improve methods of mining, not drastically, but slowly and steadily as natural and economic handicaps become greater.

B. THE ADMINISTRATION OF WORKING TIME

The non-ferrous metal mining industry in the past has been quite lax in supervision of the amount of time spent by employees at the working places. It has been shown (p. 24) that the cycle very often constituted a day's work and that all other operations of the mine were geared to the round or cycle. Since the cycle generally required a shift to complete, the miner had been paid by the shift with little thought of exactly how many actual minutes of work were contributed by the employee for his day's pay.

Many good explanations can be offered for this apparent laxness. The miner's work is too variable to supervise closely. If the round must be put in hard, tough rock, the time required for its completion, while still a shift, might be eight full working hours. On the other hand, the same miner, in the same stope, might be confronted a few days later with a weak or soft streak in which he could complete his cycle in only four hours of work. Because of ventilation conditions, however, he cannot blast his round and then go back and begin work on a second round on the same day. He must wait until the end of the shift to blast his round before he can go home. Other operations in the mine are equally indefinite.

The Fair Labor Standards Act has altered payment by the

shift. Miners now are paid by the hour and may spend only eight hours per day, plus the lunch period, under their employer's supervision. If the cycle has not been completed, its completion must be delayed until the end of the following shift.

In many mines the miner must spend one-half to two hours in traveling from the surface to the working place and return. He thus can spend only six to seven and one-half hours at the face, engaged in the completion of the cycle. Improved technology, sooner or later, will permit the completion of a satisfactory cycle in such a short time, but even then much of the time for which the miner is paid is wasted.

Much thought has been spent in the past on improvement of technology. Technological improvements are tangible, and the results of such improvement can be measured so easily that a manager can say with a fair degree of certainty that the addition of a new machine has resulted in a reduction of the work force required for the operation.

But if a change is made in the administration of working time, a great number of immeasurable factors disguise the true effects of the change. The output of the individual miner is often seriously affected by his psychological attitude. If any change is attempted that is disliked, whether the dislike is reasonable or not, the change is likely to be an economic failure. This is true whether the change is a technological one or a change in the administration of working time. But if the change is the adoption of a new machine, it can be tried first in a local area by only a few men who are chosen for their cooperativeness, so that a fair study of the economics of the change can be made. But changes in working time seldom can be made in isolated instances. Any change in working time must affect the entire working force. Suppose the change is such that ten minutes a shift that formerly was non-productive time is made productive time. Since productive time is most often less than four hundred minutes per shift at mines, a ten-minute increase should be reflected in a 2.5 percent increase of production, or production should remain constant with a 2.5 percent decrease in the size of the work force. If the work force has not been introduced to the change in the proper manner, or if the change is introduced over the objections of the union or the representatives of the employees, then the psychological attituee of the work force will be such that more more time is lost by "stalling" than is gained by the change.

Many changes in the administration of working time have been made since the inception of the Fair Labor Standards Act of 1938. The first changes that were made were changes in the length of the work week so that no overtime had to be paid. Since the law allowed forty-four hours to be worked each week without the payment of the overtime penalty for the year beginning October 24, 1938, many plans that utilized the full non-penalized time were considered and most of them tried. Forty-four hours per week can be broken down into five and one-half eight-hour days, six seven-hour and twenty-minute days or seven six-hour and seventeen-minute days.

Only a relatively few underground mines have ever found it practicable to work a half shift. The reason again is that the unit of work in mines is the cycle, and that a satisfactory economic cycle requires a shift to complete.

With the advent of the forty-two hour week, World War II already had begun and the demand for metals had risen sufficiently so that many mines were soon operating on longer weekly schedules. The few mines that had adopted a work day shorter than eight hours generally found it unsatisfactory and increased the shift to eight hours. One of the important reasons for this is that when travel time, extra lunch time, time spent looking for tools and supplies, starting and stopping, and just resting, are subtracted from the total time a mine employee is on his employer's premises, the time actually spent on productive work is far less than the time for which the man is paid. Since no accurate time studies of underground workers are available, a reasonable estimate of time paid for, but not worked, probably averages three hours per shift. If the shift is eight hours long, an average of five hours are spent in productive work. If the shift is seven hours long, only four hours are spent productively, which corresponds to a decrease in time worked of 20 percent in comparison to the 12.5 percent reduction in time paid for. If the shift is reduced to six hours in length, as it was in some mines, and the time not spent productively is again three hours, the operator only receives three hours of productive work per shift. If the time spent at productive work is no more efficient during the six-hour shift than during the eight-hour shift, the cost per unit of production on the six-hour shift is 25 percent greater than that on the eight-hour shift. For this reason and the additional reason that, in most cases, a satisfactory cycle has not been completed during the shortened shift, the six- or seven-hour shift has not proved desirable in underground metal mines.

1. Reduction in the Proportion of Non-Productive Time

The Supreme Court has decided that all time spent underground, with the exception of the lunch period, is time worked for which the employee must be paid. A large fraction of the time spent underground is non-productive time. The non-ferrous metal mining industry can thus effect a decrease in total costs if the proportion of productive time to non-productive time of the employee at work is increased. One of the longest non-productive periods spent by miners at work is spend in travel. The distance traveled underground by the miner steadily increases in any mine as that mine is extended to a greater depth and farther from the portal or collar. However, the time required to traverse this distance need not increase proportionately. Many mines worked through horizontal openings allow their men to walk to their working places. In most of these mines relatively inexpensive changes, such as the use of a man-car, could be made which would speed up the rate of travel and reduce the travel time considerably.

In mines entered by vertical or inclined openings the employee is lowered to the level on which he works by cage, car, or bucket. In mines employing a large work force, numerous trips are made before all the miners are lowered, because of the small capacity of the conveyances. Formerly a shift began at the same time for all members of the work force, but with the restrictions on the length of the shift imposed by the Act, many mines now stagger the beginning and end of each shift for the various men. That was done because some men would wait for the second, third, or fourth trip before entering the mine, and often would enter fifteen to thirty minutes after the beginning of the shift. The men who work on the same or adjacent levels are now specifically required to board a certain cage that leaves the mine collar at the time that is designated as the beginning of the shift for that group of miners. The miners who go into the mine first also come out first, and the time of arrival on the surface is eight and one-half hours after their departure from the surface. With this system the time that some men can spend at productive work has been increased as much as twenty minutes per shift.

After the miner arrives at his working level, he still may be a long distance from his working place. In the past this distance generally has been walked. With relatively small expenditures for man-car construction, and with a careful coordination of miners', trainmen's, and hoistmen's schedules, much time can be saved by transporting the miners to their working place.

A second period during which time at work is spent unproductively is just before and after the regulation lunch period. The lunch period is theoretically of one-half hour duration and is spent underground in most metal mines, though in shallow or small mines the miners prefer to spend their lunch hour on the surface. In any case, the miner seldom eats his lunch at his working place. He gathers with his confreres from neighboring working places in some dry sheltered spot, either a cool one in a hot mine or a warm one in a cool mine. Many mine operators have set up dry, warm, and clean rooms underground for the miners to eat their lunch.

In order to take advantage of the full half-hour for lunch and rest to which he feels entitled, the miner frequently stops his work as soon as he feels the boss has left the working areas. Underground bosses and supervisors generally are given the privilege of eating their lunch on the surface. Since they have other duties during the lunch period, such as making out time cards and gathering plans and data for future work, they often come up to the surface earlier and return underground later than the beginning and the end of the lunch period. The men at work know this and are extremely adroit in judging when the boss has left the area and when he will return. Consequently, the time actually spent by miners away from work probably varies from forty-five minutes to one and one-half hours.

The method usually employed to keep the men from taking such a long lunch period is to require the supervisor to stay underground until the lunch period begins and to return underground before it is over. This can only be done occasionally since the supervisor needs to use his longer lunch period to increase production in other ways. Occasional strict supervision cannot solve the problem. A more effective method, then, should be developed to reduce the lunch period to thirty minutes and effect an increase in the productive time spent by employees at work.

A third important type of non-productive time is that spent by the miner in his search for tools and supplies. Jokes crediting plumbers with arriving on the job without tools have apparently had their effect on the day's pay miner, since he often seems to arrive at his working place unprepared to work. This means that he must immediately embark on a time-consuming quest for the necessary tools and equipment. The problem of adequate tool supply is well recognized by most operators, and consequently many mines have special employees called "nippers" or tool men whose only duty is to keep each mining crew supplied with tools. There is less non-productive time spent in mines which thus keep their miners well supplied with tools.

A subsidiary problem is the supply of blasting equipment. Since a hazard exists if explosives are stored in the working place, safety rules require that blasting supplies must be gathered only immediately before their use. The trip after the dynamite, caps, and fuse consumes much time, but this time must under present technology be considered productive time. Any economical arrangement that can be made to reduce the time spent by miners gathering explosives will result in a more efficient operation.

Because of the nature of underground mining work, many other operations that require very little time on the surface, or in the factory, are time-consuming and arduous in a mine. Going to the toilet and getting a drink of water generally require separate trips to relatively distant parts of the mine. If a tool or other necessary implement is dropped, it often falls into some hole or crevice from which it is recovered with difficulty and consequent loss of time.

The time spent by the miner at work in underground mines has been shortened considerably by the provisions and the interpretations of the provisions of the Act. Since the time spent at work has been reduced, the mining industry now is faced with the problem of increasing the ratio of productive time to non-productive time through shortening the latter. Improvement in the administration of working time to reduce the proportion of productive to non-productive time will be profitable.

2. The Use of Swing Shifts and Staggered Schedules

Swing shifts and staggered schedules are used in many industries. If a plant capable of producing a certain amount of units per hour operates twenty-four hours a day, the cost of the plant, and the facilities and equipment of the plant, are amortized over three times as many units than if the plant operates only eight hours a day. A mine plant is of the type that has large investment in property and equipment and therefore one that would benefit greatly if operations could be carried on twenty-four hours a day and seven days a week. Because in most mines the cycle is completed with the blasting of the drilled holes, each mine is full of explosive gases and dust at the end of each shift. A period of at least an hour is required to dissipate and evacuate the gases and dust to such a degree that men can enter the working places and work effectively. In many mines a longer period is necessary. In any case, the time between shifts during which work cannot proceed in a mine is so long that a mine cannot operate on a three eight-hour shift basis. However, a mine can be operated effectively on two shifts per day for seven days a week. In order to do so and still not work any employees more than forty hours in any one week, swing shifts or staggered schedules must be used.

Prior to the passage of the Act, only a few mines, those which attempted to spread the work during the depression, had had experience with swing shifts. After the passage of the Act, however, swing shifts and staggered schedules were tried in many different types of mines. Though such schedules were undoubtedly troublesome to operate and did not lead to so efficient an operation as resulted from full-time employment of all crews, many operators found them cost-saving features when compared with the time and one-half penalty that otherwise would have had to be paid. Some operators, on the other hand, found that by working their employees full time and paying time and one-half for overtime, the straight schedules were cheaper than the staggered schedules.

Swing shifts and staggered schedules have many disadvantages, and apparently are not applicable to all types of mines. For one thing, they require the addition of employees to the work force. Isolated mines often have no facilities to house the additional men required. For such mines, the introduction of staggered schedules was not practicable. Very small mines also had great difficulty staggering schedules, or swinging shifts, because the number of employees engaged in any single type of operation were too few to make it possible to hire an additional man to take the place of the regular men on their days off.

Factory operations usually experience little trouble instituting a swing-shift schedule. Surface open-pit mines operate enough like manufacturing plants so that swing schedules are easily introduced and widely employed. The various types of work have been standardized and routinized so that any one of a large group of men can step in and fill a vacancy without any perceptible reduction in efficiency. A similar condition exists in most milling plants, which are on the surface and which operate almost automatically. Both open-pit mines and milling plants can and do operate on around-the-clock schedules, twenty-four hours per day, seven days per week. Swing schedules for such operations are relatively easy to set up, and a large variety are in use.

An analysis and description of the various schedules that can be and are used by open-pit mines and mills, as well as by other types of plants, was published in a recent copy of the Petroleum Engineer.[2] Figure 4-1 diagrams the Gray plan. Simply stated, the plan schedules a man to work eight hours,

FIG. 4-1. Sample shift schedule of the Gray Plan -- 42-hour week program. This plan schedules a man to work 8 hours, be off 24 hours, and repeats indefinitely without change. Each man gets one overtime day every fourth week.

ON	OFF	First Week	Second Week	Third Week	Fourth Week
		S M T W T F S	S M T W T F S	S M T W T F S	S M T W T F S
Noon	8 p.m.	A B C D A B C	D A B C D A B	C D A B C D A	B C D A B C D
8 p.m.	4 a.m.	D A B C D A B	C D A B C D A	B C D A B C D	A B C D A B C
4 a.m.	Noon	C D A B C D A*	B C D A B C D*	A B C D A B C*	D A B C D A B*

*This shift represents the sixth of "overtime" shift in this particular arrangement, which each man has one week in four.

be off twenty-four hours, and repeats indefinitely without change. Therefore, forty-two hours will be worked each week and overtime need be paid for only two hours per week. The advantages of this schedule are many. Every man has at least five full twenty-four hour periods off each week. No relief man is required. If the shifts begin at 4:00 a.m., noon, and 8:00 p.m., every man has some portion of the daytime off and thus obtains desirable sleeping time at night. The plan also allows the supervisor to see each man at least every four days and reduces overtime payments almost to the minimum.

The schedule shown in Figure 4-2 is valuable in case the constant rotation of the Gray plan is objectionable. Changing of shifts every five or six days might be preferred rather than a different shift each day. In this schedule one man works five days a week for three weeks and during the fourth week works six days. Each man therefore averages forty-two hours per week.

In the previously described schedules all shifts rotate; however, some plants allow the three regular crews to rotate but make use of a "swing shift" that remains premanently on the day shift. This is illustrated in Figure 4-3, which shows that the swing crew never works longer than forty hours a week but has the most desirable shift. The other three crews work a sixth shift each third week and therefore average 42.67 hours per week.

The problem of working a continuous week of 168 hours and yet employing men only forty hours per week is difficult if a constant crew is required. If four men are employed, 160 hours will be worked each week. The fifth man would only be needed for eight hours to complete the week. But if five sets of four men were employed, the fifth man could swing into the eight-hour vacancy in the schedules of each four-man crew, and consequently twenty-one men working forty hours a week could be used very well to fill five jobs operating a continuous 168 hours per week. Figure 4-4 shows how this schedule operates.

The schedules described previously are useful for the continuous operations of plants which offer the type of employment that can be efficiently handled by many men. Open-pit mines and mills or concentrating plants are such plants. These plants have little difficulty complying with the provisions of the Act, since the problem of scheduling the working time of employees so that the plant can operate continuously while employing men only forty hours a week is relatively simple of solution.

FIG. 4-2. A 42-hour week schedule. This plan rotates itself every six weeks. Off periods from 48 to 72 hours between shifts. Overtime weeks, time off periods, and tours equalized.

	First Week	Second Week	Third Week	Fourth Week	Fifth Week	Sixth Week
	S M T W T F S	S M T W T F S	S M T W T F S	S M T W T F S	S M T W T F S	S M T W T F S
7 a.m.–3 p.m.	A A A A A A D	D D D D D C C	C C C B B B B	B A A A A A D	D D D D D C C	C C C B B B B
3 p.m.–11 p.m.	C C B B B B B	B A A A A A D	D D D D C C C	C C B B B B B	B A A A A A A	D D D D D C C
11 p.m.–7 a.m.	D D D D C C C	C C C B B B B	B A A A A A D	D D D D C C C	C C C B B B B	B A A A A A D
A—	48 hours	40 hours	40 hours	40 hours	48 hours	40 hours
B—	40 hours	40 hours	40 hours	48 hours	40 hours	40 hours
C—	40 hours	40 hours	48 hours	40 hours	40 hours	40 hours
D—	40 hours	48 hours	40 hours	40 hours	40 hours	48 hours

FIG. 4-3. A 44-hour week schedule. This plan rotates the three regular crews and allows the "swing shift" to stand still.

	First Week	Second Week	Third Week	Fourth Week
	M T W T F S S	M T W T F S S	M T W T F S S	M T W T F S S
7 a.m. - 3 p.m.	D C C C C C D	D C C C C C D	D C C C C C D	D C C C C C D
3 p.m. - 11 p.m.	A D D A A A A	A D D A A A A	A A D A A A A	A D D A A A A
11 p.m. - 7 a.m.	B B B D D B B	B B B D B B B	B B B D D B B	B B B D B B B
	A - 40 hours	A - 40 hours	A - 48 hours	A - 40 hours
	B - 40 hours	B - 48 hours	B - 40 hours	B - 48 hours
	C - 40 hours	C - 40 hours	C - 40 hours	C - 40 hours
	D - 48 hours	D - 40 hours	D - 40 hours	D - 40 hours

FIG. 4-4. Continuous operation with employees working 40 hours per week. S.M. = swing man.

	8-4	4-12	12-8	8-4	4-12	12-8	8-4	4-12	12-8	8-4	4-12	12-8	8-4	4-12	12-8
M	A	B	C	A'	B'	D'	A''	B''	D''	A'''	C'''	D'''	A*	C*	D*
T	A	B	D	A'	B'	D'	A''	C''	D''	A'''	C'''	D'''	B*	C*	D*
W	A	B	D	A'	C'	D'	A''	C''	D''	B'''	C'''	D'''	B*	C*	S.M.
Th	A	C	D	A'	C'	D'	B''	C''	D''	B'''	C'''	S.M.	B*	C*	A*
F	A	C	D	B'	C'	D'	B''	C''	S.M.	B'''	C'''	A'''	B*	D*	A*
S	B	C	D	B'	C'	S.M.	B''	C''	A''	B'''	D'''	A'''	B*	D*	A*
S	B	C	S.M.	B'	C'	A'	B''	D''	A''	B'''	D'''	A'''	C*	D*	A*

Underground mines have been most seriously affected by
the Act and at the same time have the most difficulty with the
efficient administration of the employee's working time. It
has been shown that conditions of work underground are on a
more individualistic basis than they are in surface mines or
factories. Carter Goodrich in describing the conditions of
work of the coal miner in the 1920's described the situation
as it still is, to an important extent, in the non-ferrous metal
mines.[3]

> There is no such thing as discipline in a coal mine. The
> miner is his own boss. Neither of the two sayings is liter-
> ally true, but both are current and together they point to a
> real and important contrast between the working life inside
> the bituminous coal [substitute non-ferrous metal] mines
> and that in modern factory industries.
>
> This indiscipline of the mines is far out of line with the
> discipline of modern factories; the miner's freedom from
> supervision is at the opposite extreme from the carefully
> ordered and regimented work of the modern machine-feeder
> in a factory.[4]

Supervision in a mine often consists only of telling a new
employee where to work. His job thereafter is to complete
his cycle or his task, if he does not work on the completion
of a cycle. He generally remains in one working place for the
period he is in that company's employ or until that working
place is finished. His method of completing his daily cycle
may be different from that of any other miner in the mine.

His first move when he enters the stope is to bar down the
loose rock from the back. This he does with whatever tool he
likes best and in whatever manner he pleases. A rash man
will complete the job in a few minutes; an overly cautious
miner may dress the back for half the shift. The miner's
next move generally consists in taking a rest, perhaps light-
ing a cigarette, with the purpose of "figgerin' out" in what
order he should do his work. If the day promises to be an
easy one, he will pace himself slowly; if it promises trouble
and delay, he will "hit the ball." The blast that was fired by
the previous shift may have been perfection itself, or it may
only have shaken up the face without complete breaking. It
may have been too strong, in which case some timber may
have been blasted out, or other damage done which must be
repaired.

The miner's next move is problematical. Somehow, before

the shift is over, he will dispose of the broken rock, place the necessary timber, set-up, drill holes in such a manner that the necessary amount of rock will be broken, tear down, load the drilled holes with explosive, and then, at the end of the shift, blast. No two miners do these operations alike; in fact, they often do not do them in the same order, with the exception of blasting at the end of the shift.

The shift-boss, who is the direct supervisor of a group of underground crews, attempts to visit each working place twice daily. Any criticism of the miner's procedure is usually resented and generally not heeded. The boss's duties are to line out the work, to report progress to the foreman, to see that necessary tools, supplies, and equipment are available, to see that haulage crews keep the chutes empty, and to joke with the miner, thereby increasing the company's goodwill.

Each miner is sovereign in his own working place. This condition has made it difficult, and not very efficient, for management to attempt to double-shift working places. Stories are told of the trouble management experienced in double-shifting a rush repair job. The day shift made a good start, but the intentions of that shift were incomprehensible to the night shift who, by working hard, undid all the work of the day shift and made another good start in a different direction, but towards the same end. The funniest part of these stories is that they are often true. Miners on opposite shifts often work at cross-purposes because their methods of procedure are at variance.

Perhaps this discussion of a miner in a working place has been misleading since one man seldom works alone underground. The work unit is a crew that consists most commonly of two men, often of three men, and only rarely of four men or more. But one of these men is always dominant and he is referred to as the lead man or miner. The other men follow his lead, or take his orders. Two lead men cannot satisfactorily be worked together as a crew, and when men who are not lead men are worked together, work does not proceed at all. The real direction of operations in a mine thus comes from the miner or lead man.

A good crew is one that has learned to work together well, with each man understanding his particular duties. When a man is taken from an efficient crew, efficiency drops until the new man is broken into the style of work of the crew and the idiosyncracies of the working place.

The most efficient swing-shift or stagger schedule, therefore, must be so designed that (1) it shall not allow any man to work more than forty hours a week, (2) it shall not move a man from the working place that he has developed and to which he has become accustomed, and (3) it shall not break up an efficient work crew.

The common staggered schedule arrangement which was widely used in western mines is shown in Table 4-1. This

TABLE 4-1*. Staggered Work Schedule Used in Mines with Two-Man Crews

12 Men - 10 Jobs - 2 Rovers - 6 Day Operation - 40 Hours

Working Place	Any Shift						
	Mon.	Tues.	Wed.	Thur.	Fri.	Sat.	Sun.
1	a	f	a	a	a	a	0
	b	b	f	b	b	b	0
2	c	c	c	f	c	c	0
	d	d	d	d	f	d	0
3	e	e	e	e	e	f	0
	a'	f'	a'	a'	a'	a'	0
4	b'	b'	f'	b'	b'	b'	0
	c'	c'	c'	f'	c'	c'	0
5	d'	d'	d'	d'	f'	d'	0
	e'	e'	e'	e'	e'	f'	0

Rotation cycle completed in 1 week.
a's off Tuesday and Sunday
b's off Wednesday and Sunday
c's off Thursday and Sunday
d's off Friday and Sunday
e's off Saturday and Sunday
f's off Monday and Sunday

*From Wage and Hour Division Report.

schedule is designed for mines that operate six days a week with two-man crews while employing the miners five days weekly. Four out of each six men employed as part of the operating crews are off two non-consecutive days each week. Workmen dislike such an arrangement, since they would prefer as many non-working days to be in consecutive order as possible. Four out of six days the regular crews work together; one-third of the time, a spare man works with one of the regular crew men. This schedule loses much of its efficiency on the days that a replacement is substituted for the absent member of the crew.

Table 4-2 shows the same schedule arranged for three-man crews Two-thirds of the week one-third of the crew is replaced by a substitute. The schedule for mines that use two-man crews, operate seven days a week, and work their

TABLE 4-2. Staggered Work Schedule Used in Mines with Three-Man Crews

18 Men - 15 Jobs - 3 Rovers - 6 Day Operation - 40 Hours

Working Place	Mon.	Tues.	Wed.	Thur.	Fri.	Sat.	Sun.
1	a	x	a	a	a	a	0
	b	b	x	b	b	b	0
	c	c	c	x	c	c	0
2	d	d	d	d	x	d	0
	e	e	e	e	e	x	0
	f	y	f	f	f	f	0
3	g	g	y	g	g	g	0
	h	h	h	y	h	h	0
	i	i	i	i	y	i	0
4	j	j	j	j	j	y	0
	k	z	k	k	k	k	0
	l	l	z	l	l	l	0
5	m	m	m	z	m	m	0
	n	n	n	n	z	n	0
	p	p	p	p	p	z	0

a's, f's and k's off Tuesday and Sunday
b's, g's, and l's off Wednesday and Sunday
c's, h's, and m's off Thursday and Sunday
d's, i's, and n's off Friday and Sunday
e's, j's, and p's off Saturday and Sunday

Rovers x, y, and z off Sunday and Monday

labor five days a week is shown on Table 4-3. Four out of the seven days only one of the regular crew is in the stope, but the crew is together the remaining three days. Such disruptions of the regular crew lead to inefficiency, but if the overtime penalty were paid for the extra two days, labor costs would be increased over 14 percent. As long as the staggered schedule shown on Table 4-3 does not reduce the miner's efficiency by 14 percent, it is more profitable to apply than to pay overtime.

TABLE 4-3. Staggered Work Schedule Used in Mines with Two-Man Crews

14 Men - 10 Jobs - 4 Rovers - 7 Day Operation - 40 Hours

Working Place	Mon.	Tues.	Wed.	Thur.	Fri.	Sat.	Sun.
1	a	w	a	a	a	w	a
	b	b	w	b	b	b	w
2	x	c	c	x	c	c	c
	d	x	d	d	x	d	d
3	e	e	x	e	e	y	e
	z	f	f	w	f	f	f
4	g	y	g	g	y	g	g
	y	h	y	h	h	h	h
5	i	z	i	z	i	i	i
	j	j	z	j	z	j	j

c's and f's off Monday and Thursday
h's off Monday and Wednesday
a's off Tuesday and Saturday
d's and g's off Tuesday and Friday
i's off Tuesday and Thursday
b's off Wednesday and Sunday
e's off Wednesday and Saturday
j's off Wednesday and Friday
w's off Monday and Friday
x's and z's off Saturday and Sunday
y's off Thursday and Sunday

A schedule which has been used by some mines during World War II because of the manpower shortage is shown in Table 4-4. The mines are operated seven days each week, but work their employees six days each week, paying them time and one-half for eight hours. This schedule breaks up the regular working crew only two days in seven, but it also requires that some overtime be paid.

TABLE 4-4. Staggered Work Schedule Used in Mines with Two-Man Crews

14 Men - 12 Jobs - 2 Rovers - 7 Day Operation - 48 Hours

Working Place	Mon.	Tues.	Wed.	Thur.	Fri.	Sat.	Sun.
1	a	x	a	a	a	a	a
	b	b	x	b	b	b	b
2	c	c	c	x	c	c	c
	d	d	d	d	x	d	d
3	e	e	e	e	e	x	e
	f	f	f	f	f	f	x
4	g	y	g	g	g	g	g
	h	h	y	h	h	h	h
5	i	i	i	y	i	i	i
	j	j	j	j	y	j	j
6	k	k	k	k	k	y	k
	l	l	l	l	l	l	y

a's and g's off Tuesday
b's and h's off Wednesday
c's and i's off Thursday
d's and j's off Friday
d's and k's off Saturday
f's and l's off Sunday

Rovers x and y off Monday

Though the work schedules described in the preceding paragraphs are widely used because of their simplicity, they break rule three (a good staggered schedule does not break up an efficient work crew) too often to be truly efficient work crew) too often to be truly efficient. Unfortunately, it is impossible to measure whether the inefficiencies caused by the breaking up of a regular work crew are as costly as the overtime which is saved. That question has not yet been decided in the field, because some operators experienced with these schedules have decided opinions that staggered schedules are cheaper and others are equally certain that they are more expensive than overtime payments.

The problem, then, is to attempt to design staggered schedules devoid of most of the objections of those already described, and yet applicable to the conditions that prevail in underground metal mines. The problem is readily solved in those cases where the mine crews consist of five men. Five-man crews are rare, but often the work of the maintenance, haulage, track, and other crews can be divided among multiples of five. In this type of work, every man of the crew can perform the work of all the others, and the operations that are performed are often individual ones. For instance, the track crew tamps ties, cleans the track and the ditches, ballasts and maintains the track, changes ties, and so on. All such operations can be performed by one man or two men or twenty men working together. No standard size of crew must be maintained; little cooperation between crew members is necessary.

In situations where seven-day-per-week operation is desirable, seven men operate as a unit; two of the seven are off each day while the other five report at their working places. Each man works only five days a week, and the working place or job is operated seven days a week, as shown in Table 4-5. This "swing-shift" arrangement or a variation has often been used for those jobs that are adapted to it.

A variation employed during periods of labor shortage, in mines operating seven days per week, again employs seven men as a unit, each man six days a week. One of the seven is off each day and the six remaining men work as a crew doing the special types of work for which large five- or six-man crews are suitable. The mines that find it impossible to operate seven days and operate only six days a week, work six men as a unit and work each man five days. Each man is off Sunday and another day of the week. The five men who report for work each operational day but Sunday thus form the work crew.

Only a small portion of the work, and that generally not

TABLE 4-5. Commonly Used "Swing-Shift" Arrangement

Man No.	Mon.	Tues.	Wed.	Thur.	Fri.	Sat.	Sun.
1	ON	ON	ON	ON	ON	OFF	OFF
2	OFF	ON	ON	ON	ON	ON	OFF
3	OFF	OFF	ON	ON	ON	ON	ON
4	ON	OFF	OFF	ON	ON	ON	ON
5	ON	ON	OFF	OFF	ON	ON	ON
6	ON	ON	ON	OFF	OFF	ON	ON
7	ON	ON	ON	ON	OFF	OFF	ON

productive, can be done by five or six-man crews for which the previously described "swing-shift" arrangements are planned. Most work in underground mines requires crews consisting of two men, though three-man crews are common, and four-man crews not entirely unknown.[5]

It will be remembered that a satisfactory schedule must approach the following goals: (1) No man shall work more than forty hours a week; (2) no man shall be moved from the working place which he has developed and to which he has become accustomed; and (3) no efficient crew shall be separated.

Is it possible to set up such a schedule? The Fair Labor Standards Act requires that the overtime penalty be paid to any employee who works more than forty hours a week. This time may not be averaged. A man can be worked 160 hours in four weeks without overtime only if he does not work over forty hours in any one week. An attorney for the Department of Labor advises as follows:

> Under the presently applicable provisions of the Fair Labor Standards Act, compensation at the rate of not less than one and one-half times the regular rate of pay must be paid for all hours in excess of 40 in the work week. However, the work week of seven consecutive days may begin at any time of the day on any day in the week. Therefore, a back-to-back arrangement similar to the plan referred to in your letter might exist, but the hours of work

without overtime compensation would be limited to 40 in each week. Such a plan might be illustrated as follows: ["x" means work; "0" means no work.]

Period of Two Work Weeks

<u>S</u> <u>M</u> T <u>W</u> T F <u>S</u>: S <u>M</u> T <u>W</u> T F S
0 0 x x x x x   x x x x x 0 0

Note: It is assumed in this illustration that each day of work is an 8-hour day. Therefore, in each work week 40 hours of work were performed.

This opinion shows that if the first requirement of no overtime is rigidly enforced, the longest period that a man or a crew can be kept at work is ten days out of fourteen. A working place can be worked for a maximum of ten consecutive days by any crew without adding anyone to or separating anyone from that crew. The only way that continuous operation can be assured for any stope is either to install a swing schedule that would separate crews, or to put a different crew in a stope when the original crew is off.

To put a different crew in a stope when the original crew is off requires that the work weeks of various employees start at different times. The Department of Labor attorney offers the following opinion concerning this question: "In your letter you also ask whether there might be differing work weeks for various workers. It is quite possible that there may be varying work schedules for different groups of employees or for the various plants or departments of a large enterprise."

There are sufficient advantages to continuous operations to use swing schedules, as shown by previous discussions of such plans. It is believed, however, that the most satisfactory schedules can be worked out if whole crews are rotated through a limited number of stopes, each crew working in one of the stopes for ten days, then transferring to a second for another ten days, and so on.

The plan outlined herewith fits the condition of a mine that is in operation seven days per week, and in which the labor force conforms to the law and works but five days per week. Each crew always will mine together as a unit, whether it consists of two, three, or four men. A crew always will work opposite the same crew and will change shifts every two weeks. The men will work forty hours a week, but two weeks will be worked "back-to-back" so that each crew will work ten and lay off four consecutive days. Any crew will always begin its ten-day stretch on the same day of the week. The crew will always report to the same boss, and mine in the same region, under similar conditions. However, the crew will work in one of five working places. Seven crews will mine these five places. Two crews will always be off. Each place will be mined seven days per week, but the men will only work five days for forty hours. Table 4-6 shows how this arrangement will operate.

Crew 1 will report in working place A on Monday morning and, if the stope is operating two shifts per day, Crew 1', which always is on the opposite shift, will report in working place A on Monday evening. These crews will work in working place A for ten consecutive days. After four days of layoff, Crew 1 will report on Monday evening of the third week in working place B and work there ten consecutive days. At the beginning of the fifth week, Crew 1 will report back on day shift, but now in working place C. Crew 1 will begin on the night shift in working place D in the seventh week, on the day shift in working place E in the ninth week, and finally, in the eleventh week, return to working place A.

Crew 2 and the men on the opposite shift, Crew 2', will always follow Crew 1 and Crew 1' into every working place. Consequently, Crew 2 will always begin their week on Thursday and work ten consecutive days from Thursday through the second Saturday following. Crew 3 and Crew 3' will always follow Crew 2 and Crew 2' into every working place. They will start work on Sunday and work through until the second Tuesday following. Crew 4 follows Crew 3; Crew 5 follows Crew 4; Crew 6 follows Crew 5; Crew 7 follows Crew 6; and finally Crew 1 will begin the routine anew by following Crew 7. Inasmuch as there are seven crews and each one works ten consecutive days and the working place is operated seven days per week, then the seven crews work seventy days or ten weeks in a place before the first crew again enters it.

This plan violates the second rule, which states that no man shall be moved from the working place that he has developed and to which he has become accustomed, by moving a crew after it has spent only ten days in one stope. But that crew will return to the original stope in sixty days. During this sixty day period six other crews will have worked in the original stope and if each crew applies its individual methods of mining a loss of efficiency is incurred when they are changed. This loss of efficiency, we believe, can be corrected,

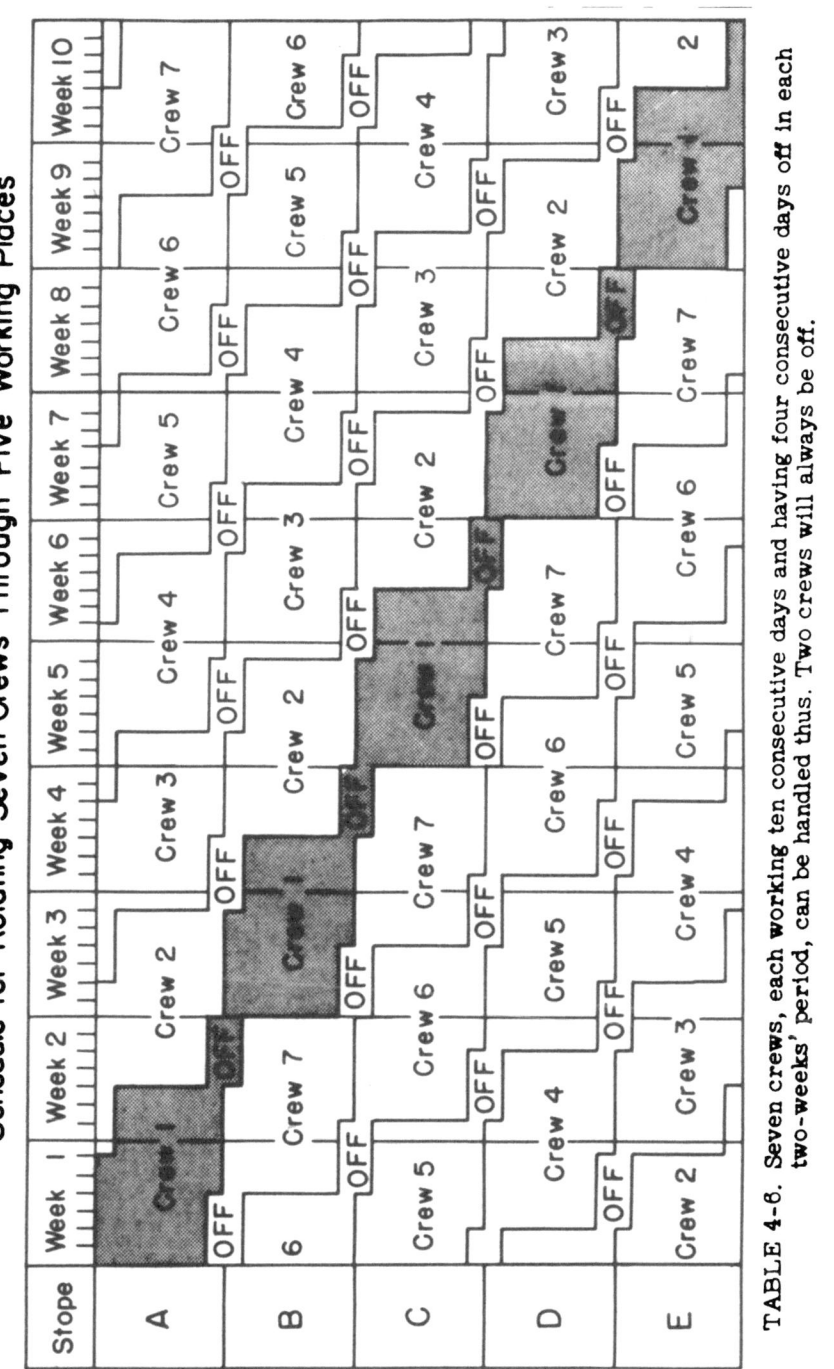

TABLE 4-8. Seven crews, each working ten consecutive days and having four consecutive days off in each two-weeks' period, can be handled thus. Two crews will always be off.

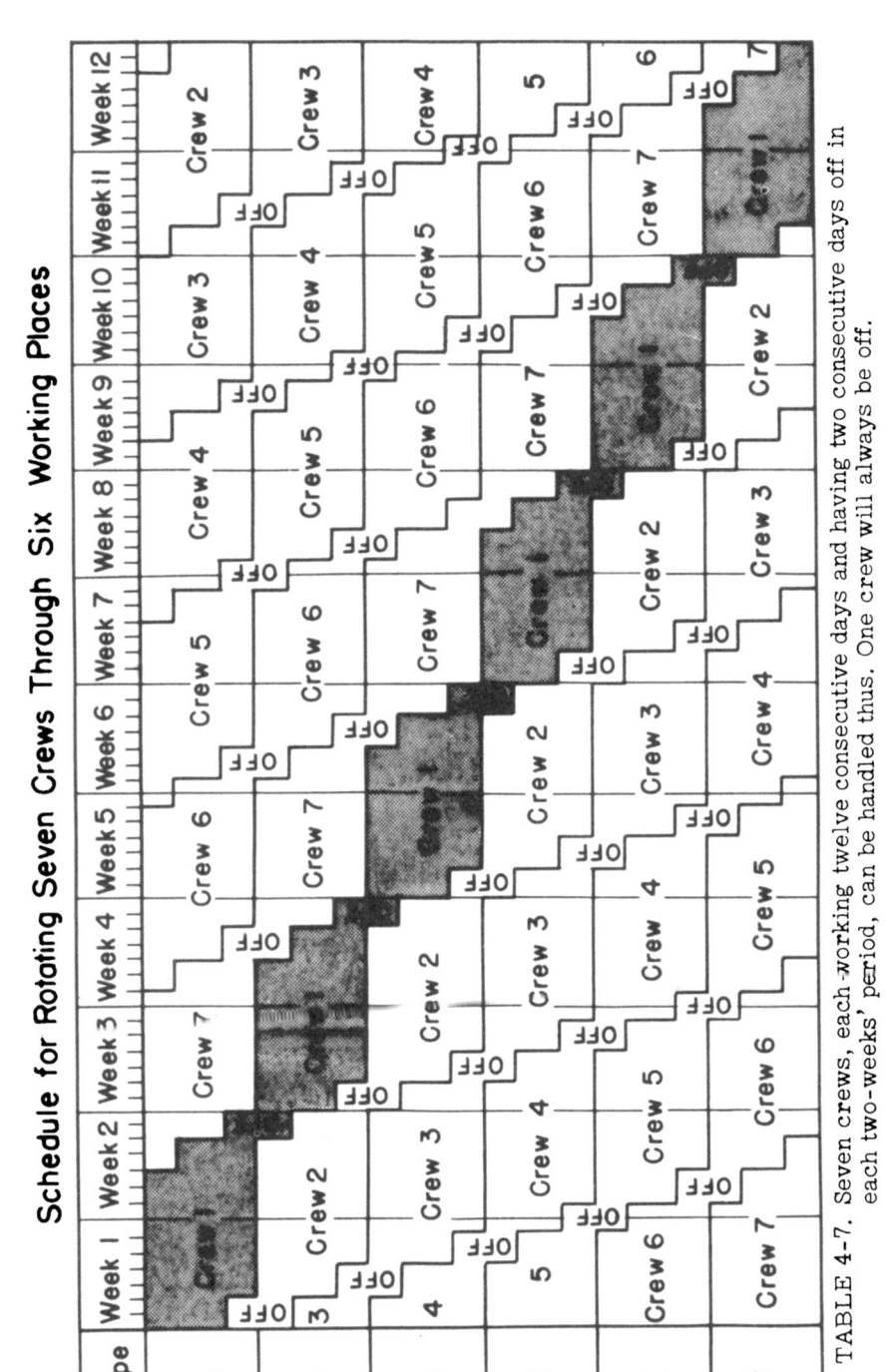

TABLE 4-7. Seven crews, each working twelve consecutive days and having two consecutive days off in each two-weeks' period, can be handled thus. One crew will always be off.

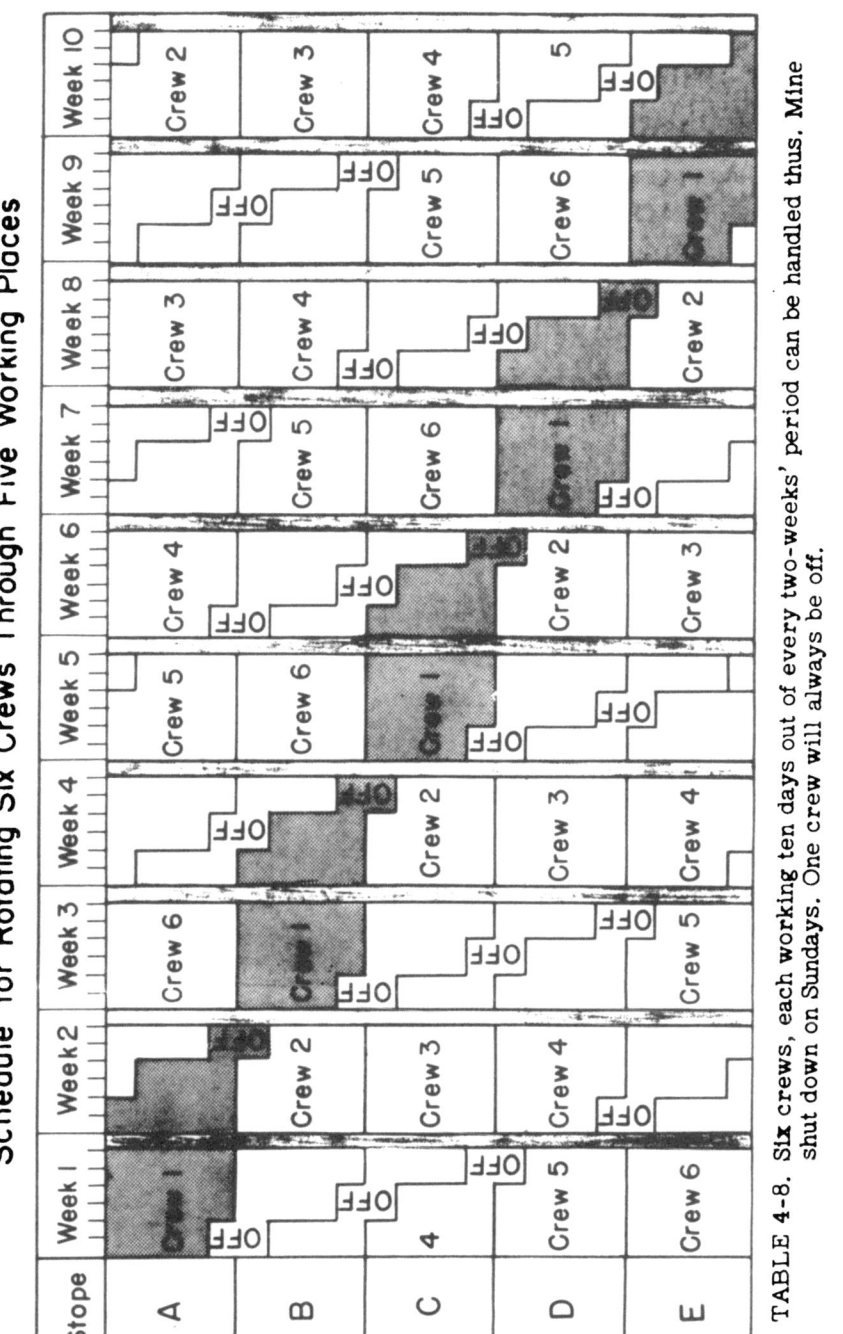

TABLE 4-8. Six crews, each working ten days out of every two-weeks' period can be handled thus. Mine shut down on Sundays. One crew will always be off.

since all seven crews work under one shift boss and spend all their time in one of five working places. Each crew always follows another crew with whose manner of procedure they soon become familiar. The end result will be that the seven crews, or fourteen crews if the working place is double-shifted, will work as a unit and the individualism of the miner will tend to become the cooperativeness of the industrial worker.

The arrangement of working time described above meets all the requirements for an efficient operation and at the same time does not require the operator to pay the overtime penalty for miners working beyond the forty-hour week. Under this plan the mine can be operated a maximum number of shifts per week with a maximum production of ore from a minimum number of working places and with a minimum amount of capital expenditure. The plan does not alter the operation of the present routine, the double shift, the change of shifts, the working day, or any of the other factors which through long usage have been established as the most efficient method of operating underground mines.

With a little modification this staggered schedule can be adapted to meet the demands of mines operating six or seven days per week while the miners work five or six days a week week. Table 4-7 shows the schedule arranged for a mine operating seven days a week at which there is such a labor shortage that the miners must be worked six days and be paid at overtime rates for eight hours each week. Each man works twelve consecutive days out of fourteen and receives two consecutive days off. Six working places must be worked as a block, but again only seven crews, with one off at all times, will work as a group. For those mines that find it impossible to work seven days a week, the schedule shown on Table 4-8 is arranged. This schedule provides for a mine operating six days a week while the miners work five days a week. All production employees are off on Sundays and two other consecutive days that are not Sundays. They work ten days in two weeks but these are never worked consecutively for every Sunday is an off-day. Again five working places will be worked as a block, but only six crews, with one off at all times, will work as a group.

Many other staggered schedules of various types that have some advantages have been devised. However, all of them have serious disadvantages, such as starting times that vary daily, or complicated changes of shifts that would cause confusion. The most practical of these appears to be the sche-

dule shown on Table 4-9. In that schedule which is designed
for two-shift, seven-day-a-week operation, three crews will
be used to double-shift one working place. These three crews
should be trained to act as a unit to obviate the difficulties
caused by changes of crews. Since each working place will
be operated 112 hours a week, the men in the three crews
will average 37 hours and 20 minutes a week. No overtime
need be paid, no crew need be broken up, and the three crews
in one working place soon would learn each other's method
of mining. Two crews will always be at work and one crew
will be off. The crews will work two stretches of nine consecutive days and then a ten-day stretch. The time off between stretches will be either 123 hours and 30 minutes or
147 hours and 30 minutes if night shift hours are the same
as day shift hours. Such a plan is suitable when unemployment is rife and can be used to spread the work among a
large group of employees when operations are curtailed,
without incurring the inefficiencies generally caused by
share-the-work plans.

The staggered schedules and swing-shift plans here presented show how thought given to the administration of working time can overcome in many underground mines some
of the increased costs that result from the Fair Labor Standards Act.

3. The Operation of Multiple Shifts

Another factor to be considered as part of the administration of working time is that of the operation of multiple
shifts. Most underground mines operate two shifts a day;
surface mines and mills frequently operate three shifts a
day. The starting and stopping times of these shifts probably
has grown up through usage from the one-shift day of the
farmer. Industrial plants at first started one-shift operation,
with the employees working from sun-up to sun-down as
they always had on the farm. As employers and society became more enlightened the length of shift gradually was reduced, first to twelve, then to eleven or ten hours a day, until now the eight-hour day is most common.

The hours of the working shift have been changed but little. The eight-hour industrial day in northern climates in
the winter is still from sun-up to sun-down, or roughly eight
in the morning until four o'clock in the afternoon. However,
modern industry long ago established multiple-shift systems
since the need to use daylight for illumination is no longer
an economic necessity. When a second shift was added, it

	Week 1	Week 2	Week 3	Week 4	Week 5	Week 6	Week 7	Week 8	Week 9	Week 10
DAY	Crew 1	Crew 2	OFF / Crew 1	Crew 3	Crew 1	Crew 3	Crew 2	Crew 3	Crew 1	OFF / 2
NIGHT	OFF / 2	Crew 3	Crew 1	OFF / Crew 2	Crew 2	Crew 3	Crew 1	OFF / Crew 2	Crew 2	Crew 3

	Week 11	Week 12	Week 13	Week 14	Week 15	Week 16	Week 17	Week 18	Week 19	Week 20
DAY	2	Crew 3	OFF / Crew 1	Crew 1	Crew 2	Crew 3	Crew 1	Crew 3	Crew 2	OFF / Crew 3
NIGHT	OFF / Crew 1	Crew 1	Crew 2	Crew 3	OFF / Crew 1	Crew 1	OFF / Crew 2	Crew 3	Crew 1	2

	Week 21	Week 22	Week 23	Week 24	Week 25	Week 26	Week 27	Week 28	Week 29	Week 30
DAY	Crew 1	Crew 2	Crew 1	Crew 3	Crew 1	Crew 3	Crew 2	Crew 3	Crew 1	OFF / 2
NIGHT	OFF / 2	Crew 3	Crew 2	OFF / Crew 3	Crew 2	Crew 3	OFF / Crew 1	OFF / Crew 2	Crew 2	Crew 3

TABLE 4-9. Schedule for rotating three crews through one working place, operating two shifts daily. Crews work nine or ten shifts consecutively, then have 123.5 or 147.5 hours off.

was added immediately after, or as close to the first shift as possible. The people on the first shift worked during the day, spent the evening at leisure, and rested at night as they always had done. But the employees on the other shifts had to disrupt their entire schedules. They were required to be at work when their bodies were accustomed to the habit of sleeping, and they were required to sleep when their bodies were in the habit of remaining awake and at work.

Naturally, the men on those shifts, other than the day shift, complained that they preferred the day shift. Management, in order to be fair to all, started the inefficient custom of changing shifts. Some industries change shifts every week, others, including the mining industry, make it a practice to change shifts every two weeks. Dr. Nathaniel Kleitman, of the Department of Physiology of the University of Chicago, points out that this rapid change of shifts leads to inefficiency, since in the second year of life the human animal has already a well-established 24-hour cycle of existence that includes work or play in the daytime and rest and sleep at night.[6] Shifts should be arranged so that this cycle is least disturbed. It has been found that an adult can, in a matter of weeks, adjust himself to a new diurnal cycle of existence so that his body will require sleep in the daytime and work at night. Since such a change does require a few weeks, if employees are required to change shifts every two weeks, their work and sleep cycles will always be upset and they will suffer constantly from sleeplessness when they are attempting to sleep and, what is worse from the employer's standpoint, they will suffer from lack of sleep and be sleepy when they are at work.

The shift schedules in most underground mines are already at their most efficient point. Work at mines begins at 7:00, 7:30, or 8:00 o'clock in the morning and between 5:00 and 7:00 o'clock in the evening. The day shift always can obtain a good night's rest at night, while the night shift which is off between 1:00 and 3:00 o'clock in the morning can also utilize much of the night for sleep.

The shifts are changed every two weeks in most non-ferrous metal mines, so the men cannot be at their peak of efficiency much of the time because of an unadjusted diurnal cycle. The solutions to this problem are multiple. One, used at present in many industrial plants, is to pay a higher wage to the men working on the less desirable shifts so that they are willing to remain on such shifts permanently. Two, rarely used as far as we are aware, is to rotate the shifts only occasionally, such as once every quarter or half-year. In both of these cases the men on night shift remain on that shift long enough to adapt their diurnal cycle to it, and consequently reach as high an efficiency on the night shift as they previously had on the day shift.

C. SUMMARY

With the increased use of technologic changes to effect improvements in mining costs, supplemented by a more rigid administration of working time, the non-ferrous metal mining industry can in time overcome the increases in mining costs that are attributable to the Fair Labor Standards Act. Though many labor-saving devices are already employed in metal mines, the adoption of technologic improvements and production-line techniques must be accelerated in order to aid in nullifying the effects of increased labor costs. In addition, through the more efficient administration of working time, an increase in the ratio of productive work to nonproductive work during the employee's working time will result. By the introduction and proper administration of well-planned swing shifts and staggered schedules, and by the scientific solution of the multiple shift problems, the employer can increase the efficiency of the miner's working time to such an extent that the hours cost, due to the shortening of the shift and week caused by the Act, will not be missed.

CONCLUSION

The non-ferrous metal mining industry is, in many respects, different from most other industries of equal importance. The differences arise from the fact that the industry mines orebodies. The orebodies are generally far from the stream of commerce. They are mined in most cases by underground methods. The price of the products of the industry are fixed either by law or by world trade factors, and the selling price cannot be changed easily to reflect changes in the domestic cost of production.

The non-ferrous metal mining industry was affected more seriously than most other industries by the passage of the Fair Labor Standards Act of 1938. The industry has costly service and overhead expenses that make it desirable for the industry to operate as many hours a week as possible. The Act forced the industry to choose between remaining on a long week and paying overtime to its employees or reducing the work week and absorbing high service and overhead costs.

In addition, the term "hours worked" in the Act was interpreted by the Supreme Court to include all hours that an employee spent underground with the exception of the lunch period. Because the mining industry was unfortunate enough to operate mostly underground mines, it was penalized more severely than other industries by the definition of "hours worked." The additional cost rise caused by paying the miners for the unproductive time that they spent traveling to and from their working places, in many cases was greater than that caused by the overtime penalty provisions of the Act.

The increase in bargaining power that the Act and other factors gave employees resulted in a rapid rise of unionism among miners. The gains that the mine employees were able to obtain by bargaining collectively with the mine operators increased the costs of mining still further, and was a greater factor than the overtime penalty and the reduction in productive hours combined.

The rise in mining cost attributable to the Act and parallel-acting factors will not be reduced by legislation to alleviate the effects of the Act. The time and one-half penalty for

CONCLUSION

overtime has been law for such a long period that it is a part of the American way of life. The "portal-to-portal" practice is now so firmly established that no factors will be able to alter it. The gains made by collective bargaining are not as permanent as the overtime penalty or the "hours worked" definition, but indications are that organized labor throughout the years will be able to wrest more concessions from the employers than they will lose to them. The era of increasing mining costs heralded by the Fair Labor Standards Act not at an end.

The ores in the ground, and the active mining industry that feeds upon them, are among the most potent resources of any highly mechanized country such as the United States. A rise in mining costs without a corresponding rise in metal prices reflects an immediate reduction in ore reserves. With higher costs, some mines that were marginal producers were forced to shut down; others were required to subtract substantial portions of their mineral deposit from the ore reserve column. Many years of improved technology and many years of increased demand must go by before the expensive re-prospecting, re-development, and rehabilitation program can be pursued that will result in the reconsideration of any abandoned mineral deposit as an ore reserve. Reduced mine production means reduced preparedness for future wars and increased dependence on foreign sources for metals. If world peace is assured, the United States may be able to afford a metal mining industry of diminishing importance; at least a metal mining industry that does not mine does not exhaust natural resources. Perhaps the best method for conserving our national resources is to reduce the size of the domestic industry and import foreign metals. However, the concensus seems to be that the sensible conservation policy is to keep a healthy, profitable mining industry which is able to continue in full production by discovering additional ore reserves as fast as the present ones are exhausted.

The metal mining industry can remain healthy and profitable only by successfully surmounting the increased costs attributable to the Act. The means of surmounting these increased costs are by the more rapid adoption of technological changes and production-line methods, and by a more rigorous and efficient administration of working time. We conclude that although the metal mining industry was stunned momentarily by the blow it received from the passage of the Fair Labor Standards Act, through intelligent and progressive action the industry will recover and attain new production records in the future.

NOTES

INTRODUCTION

1. Mining Congress Journal, Vol. 26, No. 12, (December, 1940), p. 51.
2. Fair Labor Standards Act of 1938, U.S.C.A. Tit. 29, C. 8, Section 201.
3. Memorandum to Industry Committee No. 66 for the Metal Ore, Coal, Petroleum and Natural Gas Extraction Industries, Wage and Hour Division, U.S. Dept. of Labor, (New York: September, 1943).
4. Bureau of Census, Census of Mineral Industries, (1939).
5. Hearings before the Committee on Education and Labor, United States Senate, and the Committee on Labor, House of Representatives, First Session on S. 2475 and H.R. 2700, June 15, 1947, (Washington: United States Government Printing Office, 1937), pp. 973-975.
6. Mining Congress Journal, Vol. 23, No. 12, (December, 1940), p. 34.
7. Child Labor, Children's Bureau Publication No. 197, (Washington: United States Department of Labor, 1930).
8. Ibid, (revised, 1933).
9. Population, Vol. III, The Labor Force, Sixteenth Census of the United States: 1940, (Washington: United States Government Printing Office, 1943), Table 65, p. 99.

CHAPTER I

1. Minerals Year Book, Review of 1940, (Washington: United States Government Printing Office, 1941).
2. A. B. Parsons, The Phorphyry Coppers, (New York: A.I.M.M.E., 1933), p. 9.
3. Professional Paper 127, U.S. Geological Survey, (1929), pp. 10-12.
4. Robert Peele, Mining Engineers' Handbook, (Third Edition, New York: John Wiley & Sons, 1941).
5. Waldemar Lindgren, Mineral Deposits, (New York: McGraw-Hill, 1933).

6. T. A. Rickard, Man and Metals (New York: McGraw-Hill Book Co., 1932), pp. 410-411, 579.
7. A. M. Gaudin, Principles of Mineral Dressing, (New York: McGraw-Hill, 1939), p. 1.
8. Arthur F. Taggart, Handbook of Ore Dressing, (New York: John Wiley & Sons, 1927), p. 2.
9. Minerals Yearbook, 1943, (Washington: United States Government Printing Office, 1945).

CHAPTER II

1. Frank T. DeVyver, "Regulation of Wages and Hours Prior to 1938," Law and Contemporary Problems, (1939), VI, 323.
2. J. R. Commons and J. B. Andrews, Principles of Labor Legislation, (Fourth Edition, 1936), p. 117.
3. Ibid., p. 118. Also 16 Stat. 77, (1868).
4. 34 Stat. 1415, (1907), 45 USC, Section 61, (1935).
5. 39 Stat. 721, (1916), USC, Section 65, (1935).
6. John S. Forsythe, "Legislative History of the Fair Labor Standards Act," Law and Contemporary Problems, (1939), VI, 464.
7. New York Times, (January 10, 1936), p. 6, col. 3.
8. Time, (January 11, 1937), p. 13.
9. Otto Nathan, "Favorable Economic Implication of the Fair Labor Standards Act," Law and Contemporary Problems, (1939), VI, 418.
10. Felix Frankfurter, M. M. Dawson, and J. R. Commons, State Minimum Wage Laws in Practice, National Consumers League, (1924), p. 57.
11. Hector Hetherington, "The Working of the British Trade Board System," 38 Int. Lab. Rev., (1938), pp. 478, 479.
12. Edgard Milhaud, "Results of the Adoption of the Eight-Hour Day: The Eight-Hour Day and Technical Progress," 12 Int. Lab. Rev., (1925), p. 820.

CHAPTER III

1. In discussing data from the questionnaire, the word mine includes the statistics from both mine and associated mill.
2. Harvard Law Review, (March 1941), 54: 882-4. Minnesota Law Review, (April 1941), 25: 641-2. N.Y.U. Law Quarterly Review, (May 1941), 18: 576-81.
3. International Labor Conference, Report No. 4, Vol. V: Coal Mines, (1935), pp. 31-32.
4. Coal Mines Regulation Act 1908, 8 Edw. 7, ch. 57.

5. Memorandum on Measurement of Working Time in Underground Non-ferrous Metal Mines, Research and Statistics Branch, Wage and Hour Division, (December 7, 1940).

6. Arizona Laws, (1912), c. 28, secs. 2 and 3, p. 59, Arizona Code Ann., (1939), vol. 4, sec. 56-115; Utah Laws, (1937), c. 59, amending Rev. Stat. (1933), sec. 49-3-2.

CHAPTER IV

1. John Wallace Riegel, Management, Labor and Technological Changes, (University of Michigan Press, 1942).

2. Don Attaway, "Arranging Work Schedules for Natural Gasoline Plants," Petroleum Engineer, (Dallas, Texas: November, 1943), Vol. XV, No. 2, pp. 108, 112, 114.

3. Carter Goodrich, The Miner's Freedom, (Boston: Marshall Jones Company, 1925), pp. 15-17.

4. Ibid, p. 14.

5. Calculated from information in Bulletin No. 573, Wages and Hours of Labor in Metalliferous Mines, 1924 and 1931, Bureau of Labor Statistics, (Washington: January, 1944), p. 4.

6. Nathaniel Kleitman, "A Scientific Solution of the Multiple Shift Problem," a paper presented at the seventh annual meeting of the Industrial Hygiene Foundation, Pittsburgh, November, 1942; and published in the Mining Congress Journal, (January, 1943), Vol. 29, No. 1, pp. 14, 16.

VITA

TELL ERTL

Born: 8 September 1914 at Seattle, Washington.

Degrees: B.S., College of Mines, University of Washington.
M.S., June, 1941, Columbia University.
Ph.D., June, 1946, Columbia University.

No Publications other than the present.

Bei Fragen zur Produktsicherheit wenden Sie sich bitte an:
If you have any questions regarding product safety,
please contact:

Walter de Gruyter GmbH
Genthiner Straße 13
10785 Berlin
productsafety@degruyterbrill.com